HOTWIRED

HOTWIRED

HOW THE HIDDEN POWER OF HEAT MAKES US STRONGER

BILL GIFFORD

HARPER WAVE

An Imprint of HarperCollins*Publishers*

Without limiting the exclusive rights of any author, contributor or the publisher of this publication, any unauthorized use of this publication to train generative artificial intelligence (AI) technologies is expressly prohibited. HarperCollins also exercise their rights under Article 4(3) of the Digital Single Market Directive 2019/790 and expressly reserve this publication from the text and data mining exception.

HOTWIRED. Copyright © 2026 by William C. Gifford. All rights reserved. No part of this book may be used or reproduced in any manner whatsoever without written permission except in the case of brief quotations embodied in critical articles and reviews. For information, address HarperCollins Publishers, 195 Broadway, New York, NY 10007. In Europe, HarperCollins Publishers, Macken House, 39/40 Mayor Street Upper, Dublin 1, D01 C9W8, Ireland.

HarperCollins books may be purchased for educational, business, or sales promotional use. For information, please email the Special Markets Department at SPsales@harpercollins.com.

hc.com

FIRST EDITION

Designed by Melissa Lotfy

Art on page iii by Rodrigo Corral Studio

Library of Congress Cataloging-in-Publication Data has been applied for.

ISBN 978-0-06-344802-5

Printed in the United States of America

26 27 28 29 30 LBC 5 4 3 2 1

FOR MARTHA AND ANDREW

Those who cannot be cured by medicine can be cured by surgery.
Those who cannot be cured by surgery can be cured by heat.
Those who cannot be cured by heat are to be considered incurable.
—HIPPOCRATES

CONTENTS

PREFACE › xi

INTRODUCTION › xv

I. EVOLUTION 1

1. BURNING MAN › 3
2. SWEAT › 19
3. GOOD HEAT › 39
4. TOO COOL › 57

II. ENDURANCE 69

5. ADAPTATION › 71
6. DRINKING PROBLEM › 91
7. HOTTER'N HELL › 111
8. THE NICEST KID YOU'D EVER WANT TO MEET › 121
9. TOO DARN HOT › 137

III. HEALING 153

10. STEAMED › 155

11. DETOXIFIED › 175

12. THE CASE AGAINST COLD PLUNGING › 187

13. THE HEAT CURE › 203

14. HOTWIRED › 227

ACKNOWLEDGMENTS › 239

NOTES › 243

BIBLIOGRAPHY › 263

INDEX › 285

PREFACE

Tampere, Finland
The door creaks open, and the Blister Boys are here.

One by one, they step through the low doorway, stopping first on the lower level of this bi-level sauna to rinse their naked bodies with cool water from a bucket as etiquette requires. Then they troop up the steps to the hotter upper part of the sauna and take their spots on simple plank benches. These are the hardcore sauna lovers, the guys who lined up outside the door of this establishment before it opened at two p.m. because they want to—make that *need* to—be here early, when the chamber is at its hottest and the steam is almost, but not quite, scalding.

Hence their nickname: the Blister Boys, or *rakkuloita miehet* in Finnish, a lovely but baffling tongue. They can take skin-blistering heat that would drive nearly anyone else, even the hardiest Finns, straight out.

More Blister Boys step through the door and mount the steps. Soon I am squeezed in with a dozen somewhat meaty and very naked guys happily getting their sweat on in the middle of a damp and foggy Saturday afternoon. Yet the door keeps opening, and still more pink bodies pile in and wedge themselves onto the bench, illustrating a fundamental rule of Finnish sauna: No matter how crowded it may seem, there's always room for one more. Or two.

My guide, Alex Lembke, the man in charge of heating up this hundred-year-old public sauna, is squashed into a corner between two much larger dudes so that all I can see of him are his knees and his smile. Down below, someone grabs a long-handled ladle and pours a quart of water onto the glowing rocks inside the furnace. Hot steam fills the room. Two and a half seconds later, it lashes across my back like a thousand fiery needles.

Until about a week ago, pretty much the only sauna I had ever known was the one at my gym: a dank, dark, somewhat smelly little box tucked into the back of the men's locker room, with a sign above the heater warning PLEASE DO NOT PUT WATER ON THE ROCKS. I had started going there to escape the cold Utah winter, just to put some warmth back into my bones. I grew to love the way my entire body would unclench, along with my mind, after ten minutes of stillness in the heat. But I didn't know about the steam yet.

As I had learned, shortly after arriving in Finland, that steam—which is called *löyly* in Finnish (pronounced "*low*-luh")—is the essence of sauna. The ancient Finns worshipped it, believing that the *löyly* carried holy spirits; the word itself means "spirit" or "soul."

"Without *löyly*, a sauna is dead," Alex had declared earlier that morning, as he prepared the sauna for the day. But every time someone pours water on the rocks, I brace myself.

"It hits you like a slap in the face," a friend who frequently visits Finland had warned. "You'll get used to it eventually."

Not yet. I am desperately out of my league, and the Blister Boys know it.

This place, called Rajaportti ("Rye-uh-PORT-tea"), is virtually unchanged from when it opened more than a century ago in a working-class neighborhood of Tampere, an old industrial city two hours north of Helsinki that is a little reminiscent of Pittsburgh. The sauna remains archaic in every way, from its wood-fired furnace to the rough plaster walls to the metal hooks in the changing rooms. There are no lockers, because nobody would dare steal. Men occupy one half of the sauna, women the other, separated by a door that is kept closed except on coed night, the third Thursday of every month. Entry costs eight euros, or twelve on weekends, and you can stay as long as you like.

No phones are allowed inside, ever.

Because of its old-school vibe and the quality of its *löyly*, Rajaportti is considered a bucket-list stop by sauna enthusiasts worldwide; the guestbook overflows with praise in a dozen languages, including many entries in Japanese.

The Blister Boys come every day, without fail. They crave the heat, and they can take staggering amounts of it—temperatures approaching 212 degrees F (100 degrees C). They live for the *löyly*, the hotter the better. Some are wearing hats, which seems puzzling—hats in a sauna?—but these are not like winter hats. Made of light wool or felt, these hats keep the wearer's head cool, so the rest of his or her body can handle still *more* heat.

"If you go to sauna every day, or almost every day, your threshold for coping with the heat goes up quite a bit," Alex explained. What feels unbearably hot for a normal person might seem, to the Blister Boys, "a bit chilly," he said. "These guys just need more heat—and for them, it's pleasant."

This puts Alex in a bind. As the "sauna master" of Rajaportti, he must heat it hot enough to satisfy the Blister Boys (and the Blister Girls, who can be heard laughing and chatting in the ladies' sauna next door), but not so hot that he scares off the more casual clientele, from neighborhood families to the tech-sector workers fueling Tampere's revival to sauna rookies like me. But when he fails to heat the sauna to their taste, the Blister Boys leave tart comments in the guest book.

To satisfy their heat addiction, Alex has been toiling since early in the morning, carefully stoking a fire in the base of an enormous brick-and-plaster furnace that serves both the men's and women's side of this split-sex sauna. The fire is heating up an immense pile of stones in the top half of the furnace—1.5 metric tons' worth, he says—to the point where they are glowing hot. These are special sauna stones, selected for their ability to retain enormous amounts of heat without exploding. They are changed out every six months.

After the fire had burned for a couple of hours, Alex opened the upper hatch to check on the stones inside, then consulted a red LED temperature gauge hidden in a cabinet in the ladies' changing area. It was identical to the men's side, but with children's toys piled in one corner.

"Hmmm." He frowned. "I need fifty-five more degrees."

He ducked into the wood cellar and selected a four-foot-long alder log, weighed it in his hand, then discarded it in favor of another, lighter one. The fire swallowed it whole, crackling and roaring. Satisfied, he stripped down to his underwear and began to clean both saunas, men's and women's, scrubbing down the benches with soap and bleach to prepare for the two-hundred-plus sweaty derrières they would see on this dreary March Saturday.

"Normally I work naked," he remarked. I took that as a cue to go out and find us some coffee. When I returned, he was nearly done, and it was time to enjoy the payoff for his hard work: getting into the sauna before anyone else, even ahead of the Blister Boys.

As the afternoon wears on, each hit of *löyly* becomes a tiny bit less painful than the last, and it sure beats being anywhere else on this gloomy late winter day. Alex and I wander in and out of the hot room, staying each time until it feels nearly unbearable, and then just a little longer. In between, we cool off on the benches outside, sipping nonalcoholic beers from the adjacent café, where nobody bats an eye at the sight of men and women lounging around wrapped in towels.

It begins to snow, light flakes melting into our bare skin. We head back inside, and the *löyly* greets us like an old friend. I let myself relax and enjoy the hot steam. It feels alive. So do I.

Alex nods, understanding the look on my face.

"There's no going back," he says. "You can never go back."

INTRODUCTION

> But there ain't no cure
> For the summertime blues
> —EDDIE COCHRAN

PEOPLE have strong feelings about heat. Some folks love it, like the Blister Boys or the brave souls you see out jogging or playing golf on the hottest summer days. The rest of us think those people are nuts.

Not so long ago, stepping into a scorching hot sauna would have sounded to me like one of the worst possible ideas. No thanks. I grew up loving winter and snow and skating on frozen ponds. When my family moved from upstate New York down to steamy, swampy Washington, DC, arriving in mid-August just before school began, my twelve-year-old heart sank. We were supposed to have soccer practice in *this*?

I was not alone. "August used to be a sad month for me," recalled punk-rock legend Henry Rollins, who grew up in an adjacent neighborhood. "As the days went on, the thought of school starting weighed heavily upon my young frame. That, coupled with the oppressive heat and humidity of my native Washington, DC, only seemed to heighten the misery."[1] Word up.

The only good thing about August, as Rollins noted, is that it inevitably gives way to September, when the heat finally breaks and we can breathe again.

Unfortunately, the climate appears to have other plans. Labor Day *used* to mark the start of fall; now, in many places, oppressive summer heat can drag on almost until Halloween. Ugh. In Salt Lake City, where I currently live, the high temperature topped 100 degrees F on thirty-two different days during the summer of 2022, demolishing the old record of twenty-one days—more than a month's worth of extreme heat. During July and August, daily highs typically run eight to ten degrees hotter than the long-term average.

It's much the same story around the world: Summers seem to be getting longer and hotter, while winters are steadily shrinking, to the point of nearly disappearing in some places. Scientists are quarreling over whether extreme individual weather events such as hurricanes, tornadoes, fires, and flooding can be directly attributed to climate change, but they seem to be fairly convinced that climate change is indeed causing warmer temperatures in general, and more frequent and longer heat waves, often in places that are unaccustomed to such hot weather.[2]

The media and political rhetoric around climate and weather have only made matters worse, raising our anxiety to a fever pitch, so to speak. While heat waves have been observed around the world since the beginning of civilization—a particularly dangerous one killed more than ten thousand people in and around Beijing in 1743[3]—they now come wrapped in existential dread. Weather forecasters hype any warm week in midsummer as a "deadly heat wave" caused by climate change, and caution us not to venture outside in "sauna-like" weather. As then-US Secretary of Commerce Gina Raimondo warned at the July 2022 launch of a public-health website called heat.gov, "This summer—with its oppressive and widespread heat waves—is likely to be one of the coolest summers of the rest of our lives."[4]

It wasn't, technically—the summer of 2025 was cooler, in most of the continental US*—but the vibe persists. A hot day is no longer

---

*According to data from the National Oceanic and Atmospheric Administration (NOAA). Of course, it depends where you are; it's fun to play around with the maps at https://www.ncei.noaa.gov/access/monitoring/us-maps/.

merely a hot day; it is a harbinger of our doom. As the title of a bestselling book from 2024 had it, *The Heat Will Kill You First*.

And it just might, unless violence, disease, or traffic accidents do the job first. Those heat waves are deadliest to the elderly, people with preexisting health conditions, the very poor and the unhoused, and, especially, those without access to air-conditioning (a category that includes much of the population of Europe). Also at special risk: construction workers, delivery drivers, agricultural workers, and anyone else who works outdoors, particularly in the United States, which lacks any kind of nationwide heat-safety protections for workers.

But as I would learn, heat is a double-edged sword: It can injure or even kill you, and must be approached with the utmost respect. Yet it can also heal—and even, in certain settings, make us better and stronger.

Which is how I ended up with the Blister Boys.

But I didn't expect to fall in love.

Both the Blister Boys and thermophobes (heat-haters) like me can agree on one fundamental point: Heat is uncomfortable. It can even border on painful, whether you're in a steamy Finnish sauna, a hot car on a sunny afternoon, or the London Underground on a fetid summer day. We are hardwired to understand that extreme heat can injure and potentially kill us: Our skin, gut, and brain are dotted with temperature-sensing neurons, like a zillion tiny little thermostats, to warn us when we are getting too hot.

Yet at the same time, many of us kind of like feeling a little warm. It can be comforting, relaxing, even a little sexy. Think about slipping into a hot bath after a long hike or a stressful day at work, the muscular magic that the heat seems to work, calming our bodies and minds as we sink into the warmth. No wonder our ancestors have been building and seeking out warm spaces like saunas and hot springs for thousands of years. On some deep, genetic level, we understand that the mild discomfort can be good for us. And it sure beats freezing.

Science backs this up: A remarkable series of studies out of Finland has found that frequent long-term sauna users experienced about one-half the rate of heart attacks, cardiovascular disease, and strokes over their lifetimes, and just one-third the rate of Alzheimer's disease, as those who rarely used saunas. This is an extraordinary finding, tapping into decades' worth of data to reveal that simply using sauna regularly may improve health outcomes almost as much as exercise.*

I came across the Finnish sauna studies a few years ago, while writing a book about longevity called *Outlive* with Dr. Peter Attia. Peter and I puzzled over the sauna studies with our lead researcher, Bob Kaplan, but in the end, we didn't quite know what to make of them. Sitting in a hot room for twenty minutes a day helps you live longer? According to an oddball study from Finland? We were skeptical—and suspected that there may be other factors at play—so we acknowledged the sauna studies and moved on.

A chance magazine assignment to explore the science of sweat drew me back in—and wound up completely changing my mind about heat. After a deeper dive into the sauna studies, some long conversations with experts, and some even longer (and far more relaxing) sauna sessions around the world—from Helsinki to Hong Kong, New York to Tampa, the Liverpool docks to the woods of Northern Minnesota, and of course with the Blister Boys—I became convinced of the power of heat to improve our health and enhance our lives.

Most persuasive, to me, is the fact that so many different cultures around the world participate in similar heat-bathing rituals, from Native American sweat lodges to the Mexican *temescal*, the Russian *banya*, the Middle Eastern *hammam*, the Korean *jjimjilbang*, and the Japanese *onsen*, hot baths fed by volcanic springs; there are even Mongolian and Irish versions of the heat bath. All represent variations on the same basic recipe: Heat an enclosed space to a ridiculously high temperature using some combination of fire, rocks, and water, and then add people. Steam may be optional, but sweating is mandatory.

"There are sweat baths all over the world; it's as common as the baking of bread or fermenting the grape into wine," says Mikkel Aa-

---

* Almost, but perhaps not quite: Exercise has been studied far more thoroughly than sauna use, and while their cardiovascular effects are similar, exercise also (obviously) requires the activation of muscle, which brings an additional suite of benefits, in areas ranging from metabolism to brain health. Thus exercise is a more powerful stimulus than sauna alone. See Feldman, D. I., Al-Mallah, M. H., Keteyian, S. J., Brawner, C. A., Feldman, T., Blumenthal, R. S. & Blaha, M. J. (2015). "No evidence of an upper threshold for mortality benefit at high levels of cardiorespiratory fitness." *Journal of the American College of Cardiology* 65(6), 629–630. https://doi.org/10.1016/j.jacc.2014.11.030.

land, author of *Sweat*, a study of global heat-bathing traditions originally published in 1978. "It's a human phenomenon, to go into a hot room and sweat."[5]

This, too, may be hardwired into our biology, because sweating is something that we humans do exceptionally well. This is not an accident. Our ability to sweat to cool ourselves may be one of our evolutionary superpowers, a trait that enabled our long-ago ancestors to climb from the middle of the food chain all the way to the very top.

The surprising thing is that sweating may *still* be giving us a leg up in ways that we have almost forgotten about: by improving our physical health and making us better athletes, for starters, but also helping to forge social connections and build our physical and emotional resilience. Heat exposure even makes us stronger at the cellular level. Getting a little bit hot and sweaty can help make us better humans.

And our world is definitely getting hotter and sweatier.

Yet despite the dire warnings from TV weather forecasters, we human beings have an amazing capacity to cope with a hot day—to "thermoregulate," as the scientists put it—that is far superior to that of any other animal. Our ancestors evolved in the African heat, and our ability to sweat is at least a million years old. "It's hot now, but it was hotter then," at times, points out Daniel Lieberman, a professor of evolutionary biology at Harvard and author of, among other books, *The Story of the Human Body*.

Not only that, but we humans are endowed with the ability to adapt to *both* heat and cold, from humid jungles to Arctic tundra to hot, dry deserts. You already know this: That 80-degree (Fahrenheit) afternoon that feels unbearable in May will seem like a blessed relief by August. The reason is because over the summer you become *heat-adapted*, which is yet another human superpower. Olympic athletes are now exploiting this superpower to gain an edge over their rivals, breaking through old barriers and achieving new levels of performance. The rest of us can do the same, as I learned quite vividly on a 100-mile bike ride in Texas heat that should have landed me in the ER. Sweat isn't merely a drippy, sticky, sometimes uncomfortable by-product of exercise; it is a kind of performance-enhancing drug.

The most unexpected thing I learned about heat is its power to heal.

For as long as we humans have known fire, we've sought out hot spaces to clean and restore our bodies, to heal from disease, to mingle with others (sometimes to get laid), and to preserve our emotional equilibrium. Medieval physicians often plied their trade in saunas and bathhouses, long before anyone knew that the heat killed germs. (Or even that "germs" existed.) More than that, institutions such as the sauna and the sweat lodge (and the *temescal, onsen, jjimjjilbang, banya, hammam,* and even the ancient Roman baths) also served as spaces for community-building, ritual, gossip, and social connection, an antidote to the isolation that plagues so many modern lives.

Even so, it is tempting, even natural, to want to avoid heat and hot weather, or any situation that forces us to get sweaty—to hide in the air-conditioning, if we are lucky enough to have it. But as I've learned over the last two years, the answer is sometimes the opposite of that—to lean into the heat. Just a few decades ago, our grandparents and great-grandparents would have experienced big temperature swings, from cold to hot to cold again, in the course of a single day. Now many of us have engineered that out of our daily lives, living instead at a constant seventyish degrees Fahrenheit.

It's similar to exercise. Over the last few decades, we have systematically removed the need for physical activity from our lives. Instead of walking anywhere or lifting anything, we drive to a place called "the gym" to walk on a treadmill and pull and push on machines that mimic the movements that our ancestors needed to perform merely to survive. Just as we benefit from exercising our muscles, I would argue, we also need to "exercise" our sweating machinery—and everything that goes along with it.

For many of us, that machinery has gotten pretty rusty. Living in climate-controlled comfort, we have all but forgotten how to sweat, let alone how to tolerate temperatures outside the magical modern comfort zone. The more time we spend inside, the less comfortable we feel outside. It requires a conscious effort to go into the heat, and to relearn the thermal flexibility that our ancestors knew. While it is not always comfortable, the payoff is worthwhile. Just one teaser: Spending time in a hot sauna, just sitting there doing nothing but sweating, is functionally similar—in some ways—to going for a light jog. (Even better: Do the light jog, *then* sit in the sauna.)

My day with the Blister Boys was only one stop on a two-year exploration of the power of heat. I have dug into the science of heat adaptation with experts in the field and talked to athletes, workers, scientists, sauna addicts, and people like the Blister Boys who have made heat exposure an integral part of their lives. Like CJ Albertson, a substitute teacher and former college runner who rigged chicken-coop heat lamps around his basement treadmill and used heat training to become one of the best American marathoners of his generation. Or Ryan Swoboda, a former college football player who nearly died of heatstroke in preseason practice—but who used the science of heat adaptation to recover from his injuries and reach the NFL draft.

At first, these thermal explorers seemed a little eccentric. But soon enough I had become one of them. I sweated all over the world: in Finland, with the Blister Boys and hundreds of other sauna-goers; in Connecticut, at a cutting-edge heat lab dedicated to pushing the limits of "heat adaptation" for athletes and to protecting workers from heatstroke; in Texas, joining several thousand other nutjobs to cycle 100 miles on one of the hottest days in one of the hottest years ever recorded there; in New York City, at a hybrid sauna/nightclub/meditation center where I shared personal stories with perfect strangers; closer to home, where I schvitzed all over the floor of my local hot yoga studio; in Colorado, testing a groundbreaking mental-health therapy combining an infrared sauna and a cold plunge as a potential treatment for depression; and in Bucharest, Romania, where I found bliss at a massive thermal bathing park and sauna complex that is like a modern reincarnation of the ancient Roman baths, with a dash of Disney World.

Through my journey, I became convinced, by both the science and my own experiences, that heat adaptation, heat training, and heat-based therapies might just help us tap into an evolutionary advantage that most of us have forgotten that we even possess.[1] So think of this book as not merely a survival handbook for climate change but a guide to living a fuller, more contented life by unlocking, or rediscovering, our superpower to sweat.

Whether you think you hate the heat, like I once did, or love it, like I do now, it is hard to deny: We were born to sweat. And sweating makes us stronger.

I
EVOLUTION

1

BURNING MAN

> If I owned Texas and Hell, I'd rent out
> Texas and live at the other place.
> —GENERAL PHILIP SHERIDAN, 1866

A HUGE white pickup truck, the kind with dual rear wheels, roars by, pulling a flatbed trailer loaded with bodies. A few minutes later, another rumbles past. Then another. Each trailer carries a pile of sad cyclists, slumped in defeat, too exhausted to take off their helmets as they are hauled back to the starting line with their bikes. Their faces are pale and vacant, their clothes crusty with salt from their sweat.

I keep on pedaling, wondering if I too will end up on the trailer of doom. A few miles farther on, I swerve around more bodies sprawled on the tarmac. My heart leaps—was there some kind of accident? Nope, just more beat-up bike riders resting in the shade of a spreading live oak, too wiped out to care that they are napping on a freeway frontage road.

A bank thermometer informs all and sundry that it is 107 degrees F out, information that nobody wants or needs right now. The blacktop beneath is so hot that it feels angry. A sign outside a rundown church advises:

SIN-BAD
JESUS-GOOD
DETAILS-INSIDE

And below that, an apology:

TOO HOT TO CHANGE SIGN

It's too hot to do anything. Definitely too hot to be riding a bike outside in the sun. I take a sip of lukewarm water from my bottle, squirt some more down the back of my jersey, and pray that my tires don't explode from the heat.

Two riders ahead of me pull off the road and duck under the awning of an auto parts shop, fleeing the merciless sun. I dare not join them, lest I too end up in a "sag wagon." So I keep turning the pedals of my rented bike, as I've been doing for the last four hours, as if through a narcotic dream.

The words of a Texan acquaintance loop in my brain: "You'll never make it," he had assured me weeks earlier, "'cause you ain't from there."

Five months before, I was noodling on my phone after yet another session in the sad gym sauna when up popped the website for this bike ride, which is aptly known as the "Hotter'N Hell Hundred." One hundred miles, at (roughly) one hundred degrees Fahrenheit. Held every August on the roads around Wichita Falls, a small city of 100,000 about two hours north of Dallas,[*] the Hotter'N Hell is a Texas institution that exists for one reason only: the heat. That sounded vaguely appealing in snowy March. A few clicks later, I was committed.

I didn't know then that the summer of 2023 would turn out to be the Hottest Summer in History, with heat waves sweeping the globe. Nor had I fully considered the implications of riding a bicycle for a hundred miles in hundred-degree-Fahrenheit weather, wearing overly tight black cycling shorts. I was just bored and looking for a challenge.

---

[*] Wichita Falls is also home to the Professional Wrestling Hall of Fame and Museum.

I got one.

A decade or so ago, riding my bike for 100 miles would have been no big deal. I was then what people call an "avid" cyclist—rhymes with "rabid." But that was ten years and a similar number of pounds ago. These days, my go-to route is twelve miles long. And I would never, ever ride in hot weather.

I signed up for the Hotter'N Hell without really thinking it through, is what I'm saying. And by the looks of it, a few of the nine thousand other riders hadn't either. All morning long, we'd been steadily heating up, like the proverbial frogs in boiling water, to the point where some of us were beginning to get wobbly.

"If you're feeling like this already," a fortyish man had warned his gray-haired, gray-faced, somewhat unsteady dad, who looked to be about seventy, at one of the rest stops, "you're not going to feel any better in thirty miles."

Honestly, they both looked thoroughly poached. And when I checked myself in my phone camera, so did I. Who was this red-faced guy staring back at me with wild eyes? What was he doing here?

Dreamed up by members of the local bike club back in 1982 to mark the centennial of Wichita Falls, and also to bring some much-needed visitors to town, the Hotter'N Hell Hundred probably shouldn't have been run more than once or twice. Who would pay money to come to Wichita Falls in August to ride for hours on shadeless, broiling-hot North Texas farm roads? Let alone do it again?

Lots of people, it turned out. Despite—or perhaps because of—the fact that it is so off-the-charts extreme, the Hotter'N Hell caught on, growing into the largest organized "century" bike ride (a single-day ride of 100 miles) in the US. Some years, as many as fourteen thousand riders have shown up, from all over the world. Because simply cycling 100 miles in Texas summer heat apparently isn't enough for some folks, the Hotter'N Hell has expanded into a four-day "Festival of Fitness," with multiple bike rides and races, including mountain biking and even—perish the thought—trail runs. The only event that sounds remotely appealing right now is a one-mile "bar crawl" that takes place later tonight.

We certainly aren't here for the scenery. The roads run dead straight across a flat, desolate scrubland dotted with rusted-out pump jacks and a string of busted boomtowns. This area was the heart of the fabled Texas oil patch that, ironically, disgorged a huge quantity of the fossil fuels that helped get climate change going back in the first half of the twentieth century. Now we are reaping the consequences in the form of abnormally warm temperatures across the globe—and on these lonely chip-and-seal country roads.

But this is exactly what we came for. The Hotter'N Hell is a rite of passage for residents of the Lone Star State, who make up the vast majority of entrants. The heat is part of their identity. Many of the riders appear to be Texans first and athletes second. They are a diverse lot, all ages and shapes and sizes, and definitely not all fitting the stereotype of shrink-wrapped, Lycra-clad, ultra-lean Type A road cyclists. We've got white-haired grandmas, barbecue dads, girlfriends' groups, families with teenagers, and everyone in between. As different as we all are, we share two common goals: to make it to the finish, and to prove that we are capable of doing way more than anyone else thinks.

"I had thought it would be fun," wrote author Lawrence Wright in *Texas Monthly* magazine after completing the 1989 edition of the Hotter'N Hell—one of the hottest on record, with high temperatures hitting 110 degrees F on the road.

"But that's not the word I would use now. It was insane."

A few months before the ride, I opened Google and typed, "Hotter'N Hell Hundred—"

To which the algorithm helpfully added *"deaths."*

Further research revealed that there had only been a handful of fatalities in the event's forty-four-year history, a fact that I chose to view as comforting—coffin half empty, as opposed to coffin half full. The year Wright rode it, in 1989, ninety-two riders had ended up in the ER, and one had died of a heart attack. That was a bad year. Now the organizers field a large and attentive medical team, with more than a thousand volunteers including doctors, nurses, and EMTs spread along the route, and rest stops with large shady tents and cots and ice tubs every ten or twelve miles. There would also be an air ambulance standing by just in case, a fact that I found both reassuring and alarming.

Some years, hundreds of people have required IVs or other treatment on the road, but serious incidents have been relatively few. The fact that most riders come from Texas and are thus at least somewhat accustomed to the heat has surely helped to keep the body count low. My Texan friend had said that his mom had even done it in her early sixties. I, on the other hand, was "not from there." So I would need to heed his mom's advice: "Prior planning," she liked to say, "prevents piss-poor performance."

In the interest of preventing piss-poor performance, not to mention a ride in the air ambulance, I reached out to Robert Huggins, an exercise physiologist and heat expert at the Korey Stringer Institute (KSI) at the University of Connecticut. Named for a beloved Minnesota Vikings lineman who had died of heatstroke after an early-season football practice in August 2001, the Korey Stringer Institute is one of the leading heat-physiology laboratories in the world. It conducts research into heat and its effects on health and athletic performance, runs public-education campaigns on heat safety for athletes and for companies that employ outdoor workers in hot places, and, lastly, it lobbies for heat-protection measures at the state and federal level to prevent athletes and workers and regular people from dying of heat illnesses.

KSI also offers heat training and heat coaching to professional and elite amateur athletes. I hoped to enlist their help in preparing me, a decidedly unprofessional and very sub-elite athlete, for the ride. But my pitch did not go as planned. When I told Huggins that I had already signed up for the Hotter'N Hell, he exclaimed, "Oh, you're crazy to do that!"

Before he could click "LEAVE MEETING," I managed to blurt out, "*You've got to help me!*"

He sighed.

Huggins had been to the Hotter'N Hell as part of KSI research teams, observing the effects of severe heat and extreme distance on athletes' bodies. Apparently, the ride was kind of a laboratory for studying lunatics who took on challenges that maybe they should have skipped. It was never pretty.

One year, Huggins said, he was stationed at a rest stop at Mile 86, the hottest part of the course and often the point where riders had reached (or exceeded) their absolute limit. The temperature had climbed to 109 degrees F that day. His job was to collect data on a small group of riders who had volunteered to be study subjects, but he was horrified at the carnage going on around him. "I watched people who were *having heatstrokes* get back on their bikes and keep riding," he told me incredulously. "And the medical staff didn't stop them!"

As an outside researcher and not a medical volunteer, he was not permitted to interfere. Years later, he remained slightly traumatized by the experience. He shook his head. If I was going to have any hope of surviving the Hotter'N Hell, he said, I needed to get myself to Connecticut ASAP.

In late June 2023, I flew to Hartford and found my way to the UConn campus, set amid stone-walled farms and low wooded hills. It took some sleuthing to locate the KSI heat lab, tucked into the bowels of the UConn Huskies' famed basketball arena; both the men's and women's teams have won multiple national championships.

The heat lab is a bit less glamorous, a cramped space with windows facing a hallway. Huggins was waiting for me there, along with two graduate-student assistants, Sean Langan and David Martin. The latter two are lean, lanky triathlete types—a common build in the world of sports science—while Huggins, a dad of two in his late thirties, is more solid. His colleagues call him "Huggy Bear," and he was sporting a fat lip from a hockey game the previous evening.

Had I known how the next two hours would go, I would have gladly taken an elbow to the mouth instead.

Huggy Bear opened the door to the heat chamber, and a blast of hot air clobbered me in the face. The 440-square-foot chamber had metal walls and thick glass windows, like an oven, and a powerful heating and ventilation system capable of warming (or cooling) the chamber to any temperature between 32 and 110 degrees Fahrenheit, with humidity from zero to 90 percent. In addition to testing athletes, the lab also conducts studies for the Department of Defense—which is extremely interested in the effects of heat on its soldiers, for obvious reasons—as well as for private companies gauging the effects of heat on workers.

"This thing never shuts off," Huggy Bear said over the roar of the fans. Today, he had it set at 93 degrees F (34 degrees C) and 38 percent humidity. Nasty.

I tried to remind myself that a growing number of people around the world *live* in such conditions, or worse, in places like Pakistan, the Arabian Peninsula, Mexico, and Southeast Asia. (Also, at times, my native Washington, DC.) They essentially live in a heat chamber. This constant heat stress, without the ubiquitous air-conditioning to which many Americans are accustomed, puts the poor and elderly at great risk; it is an emerging global health threat.

I was here by choice.

Soon I had changed into my bike clothes and was pedaling away on a stationary bicycle positioned inside the lab. This bike had seen better days: The water bottle cage was broken and bent, its jagged metal end dangling perilously close to my right calf muscle as I pedaled. The racing-style handlebars were turned upward, in a position that cyclists call "DUI-style," because it is commonly seen on bikes pedaled by disheveled middle-aged men with one too many drunk-driving violations. It looked like Satan's personal Peloton.

Despite its janky appearance, this bike was a highly sophisticated piece of equipment called a Velotron, which is often used in scientific research. Its computer-controlled flywheel, a solid copper disc that spun like a deli slicer's blade, could be calibrated to a precise level of resistance. Huggins set the power level to 140 watts, an easy cruising pace for a fit cyclist—which, I quickly learned, I was not. Fitness-wise, I was maybe closer to the DUI guys. Like them, I soon began to regret the choices that had put me on this dreadful contraption. Within minutes, sweat began dribbling from the end of my nose onto the frame of the bike, drop by drop. A fan whirred in my face to simulate the airflow of riding a real bike, but it only made me feel hotter.

In short order, I was panting for breath, my face the color of raw hamburger meat. Quietly, Huggy Bear cranked the heat up to 102 degrees F with 43 percent humidity. Research from Penn State suggests that these conditions are close to the upper limit of what humans can tolerate, or what scientists call "uncompensable"—meaning that most people will struggle to maintain a stable body temperature, and many will fail.[1]

To riders in the Hotter'N Hell Hundred, on the other hand, it would qualify as a pretty nice day.

The hot environment was only one of my problems. Most of the heat that I was experiencing was coming from inside my own body, Huggins later explained. A down-to-earth guy from outside New Haven, he likes to compare the human body to a car. In the same way that an automobile engine will get hot as it propels the vehicle forward, our bodies generate a huge amount of excess heat energy when we do anything physical. Whether we are running or cycling or walking or doing other physical work, only about 20 percent of the energy that we burn in our muscles is translated into mechanical work and physical motion—which is, surprisingly, about the same level of efficiency as an internal combustion engine.

The rest takes the form of heat. Scientists call this "metabolic heat," while normal people know it as *body heat*. As I worked to satisfy the Velotron's demand for 140 watts of power, my muscles were generating an extra 560 watts' (or so) worth of heat energy that, quite frankly, I neither wanted nor needed. It was as though I had swallowed a small hair dryer, and it was heating me up from within.

I considered bolting from the hot room and diving into the deep-blue fifty-yard competition pool just across the hall. But the pool door was locked. More relevantly, I was tethered in place to a stack of blinking, beeping monitors by a long gray cord that led up and over the bike's handlebars and snaked down into my bike shorts, ending in a temperature probe lodged firmly in my hindquarters. Escape was not an option.

On the phone, Huggins had promised that they would monitor my core temperature using a more modern, less invasive method, a high-tech pill-sized capsule that I would swallow. It would then transmit temperature data electronically as it made its way harmlessly through my digestive tract. But someone had forgotten to charge the capsules overnight, supposedly, so I was dispatched to the lavatory with the rectal probe, a jar of Vaseline, and some perfunctory instructions.

"If it comes out your mouth," Huggins had said, helpfully, "it's in too far."

In the world of people who study physiology and exercise science, the rectal probe is just a fact of life; everyone you meet has experienced it at least once, and usually many times. It also happens to be the gold standard for taking core body temperature readings and has been a staple of physiology research for a century. So there was no getting out of it. It made sitting on the bike seat a little awkward, but soon enough I got used to my new friend, who I called "Probey." And I could refocus on not dying.

Every few minutes, a cheerful lab intern named Nia McBride held a printed card in front of my face, asking me to rate my "perceived exertion" on a scale of six to twenty. How hard did it feel like I was exercising?

I lowballed her, hoping to fool myself into thinking that I felt better than I did. It didn't work. *Somewhat hard* became *very hard*, verging on *extremely hard*. Meanwhile, Probey was telling his own sorry tale. As I pedaled in the hot room, going nowhere, my core body temperature crept steadily upward, a hundredth of a degree at a time, up and up and up. My energy flagged. I was going the equivalent of ten miles an hour, which is not fast.

"Imagine you're at mile eighty-eight of the Hotter'N Hell!" Huggy Bear said. "You've got this!"

I groaned. He looked at me with pity.

"I'm not the Devil," he said finally. "But I'm not making it any cooler for you. It's going to be hotter in Texas."

He was not wrong.

As painful and absurd as it seemed, in the moment, my session in the KSI heat chamber was a standard version of something called a "heat tolerance test." Similar protocols have been used by the US and Israeli militaries,* as well as in scientific research, for decades. The basic idea: Put someone in extremely hot conditions, make them do something

---

* In the classic Israeli Defense Forces heat tolerance test, subjects walk for two hours on a treadmill at 3 mph (5 km/hour) at a 2 percent grade in a temperature of 104 degrees F (40 degrees C) and 40 percent relative humidity. You fail the test if your core temperature exceeds 101.3 degrees F (38.5 degrees C) and/or your heart rate exceeds 150 bpm. Good luck with that. However, many different versions of heat tolerance testing exist, tailored to specific activities and conditions.

physically demanding (typically running or walking on a treadmill) with a temperature probe in their backside, and see what happens.

The KSI lab is a state-of-the-art facility, but the idea of heat testing originated nearly a century ago in an unlikely place: the gold mines of South Africa.[2] Gold had been discovered near present-day Johannesburg in the 1880s, and as the mines pushed deeper into the earth, the tunnels got hotter and hotter. Because the miners sprayed water to keep down the dust, the relative humidity underground approached 100 percent, making conditions even more oppressive.

The white mine supervisors had assumed that the workers, who were all native Africans, were somehow impervious to heat. But when mine workers began dying from heatstroke, bringing gold production to a halt, they realized that they had a problem.[3] After twenty-seven miners died in 1930 alone, the mine owners assigned a young medical officer named Dr. Aldo Dreosti to study the problem and try to find a way to keep the miners alive and mining.

Clearly, some workers did just fine in the heat, while others did not. To tell them apart, Dreosti devised a simple test that was not all that different from what KSI and many other laboratories do today. Inside the mine hospital, Dreosti set up special tents that were heated to 95 degrees F. New workers were brought into the tents and told to shovel rocks for an hour, as though they were working in the mine. The doctor and his nursing staff would monitor their heart rate, their blood pressure, how much they sweated, and, especially, their body temperature.

It turned out that the change in a worker's body temperature over the course of the one-hour test was the *only* factor that predicted how well he would fare in the heat of the mines. If his rectal temperature stayed below 100.6 degrees F while working in the heat for an hour, a worker was classified as "heat tolerant" and okayed to work after undergoing some basic training sessions. If his body temperature exceeded 102 degrees F (38.9 degrees C), he was flagged as "heat intolerant" and prescribed up to fourteen days of heat training. These heat-training sessions were almost identical to the heat tolerance test: an hour of shoveling rocks in a hot tent. After four or five sessions (or more in some cases), the workers would grow accustomed to laboring in the heat.

Dreosti tested tens of thousands of workers from all over Africa, and found that, contrary to common belief, not everyone with black skin was magically impervious to heat. The capacity to handle hot conditions varied widely among individuals, even those of similar genetic or regional backgrounds. Some 15 percent of the new recruits were found to be heat intolerant and thus potentially at risk—unless they were trained to adapt to the heat.

Under this new testing and training regime, many fewer workers died, and gold production soared. The mine owners' association awarded Dr. Dreosti a medal for his efforts. His work had not only saved lives but also helped to lay the foundation for much of modern sports science. The basic idea was easy and straightforward, a simple protocol of testing and training. But the underlying principle was far more profound. With the miners, a natural experiment involving tens of thousands of subjects, Dreosti had shown that the limits of human endurance in the heat are incredibly elastic; we can do more than we think.

I would not have fared well in the gold mines of the Witwatersrand. The longer I pedaled the hateful Velotron bike with its broken water bottle cage and DUI-style handlebars, the worse I felt. The pedals seemed to grow heavier with every turn, and my body temperature kept climbing. My heart rate also accelerated, struggling to supply my muscles with oxygen while also keeping me cool. This is called "cardiovascular drift" or "CV drift," where heart rate climbs ever higher despite a constant workload. Its exact causes are not completely known, but CV drift is common in people who are unacclimated to heat. As it continues, an athlete's effort will steadily diminish to the point where they simply must stop.

Even when I finished the first part of the test and rested in a chair for ten minutes, my temperature kept on rising inexorably. It was like global warming was taking place inside my body. And I wasn't done yet; next, Huggy Bear made me ride a "time trial," going as fast as possible on the stationary bike for fifteen minutes. Just what I didn't want to do. By the time *that* was over with, my core temperature had soared all the way up to 102.2 degrees F (39 degrees C). At 104 degrees F (40 degrees C), just one degree C hotter, research safety guidelines would have required Huggins to pull me from the room and stop the test. (Which I would have welcomed.)

Later, in his windowless office under the arena stands, he squinted at his computer screen and frowned. "You didn't really show any signs of a plateau, in heart rate or core temperature," he said, shaking his head. That meant that my body never reached thermal equilibrium, or homeostasis. My core temperature had just kept going up. In fact, I was *still* sweating all over his office chair. Had I continued, he added, "something would have had to give."

He was telling me, politely but directly, that I had resoundingly flunked the heat tolerance test. And the Hotter'N Hell Hundred was barely two months away.

I hadn't traveled all the way to Connecticut to find out that I don't do well in the heat. I already knew that. I had come seeking answers, a solution to my problem. And I already knew what the solution would be, thanks to an extraordinarily strong and lucky man named Ryan Swoboda.

In the summer of 2017, Swoboda showed up for football practice at the University of Virginia as a freshman recruit. Held in the afternoon heat, these tough preseason workouts were intended to whip the rookies into shape and also to help cull the weak. The torture of choice was an exercise called "up-downs," where the players jog in place, drop to the ground, and then get back up and keep jogging. Exhausting.

Up-downs might not be a big deal for a six-foot-two, 200-pound quarterback, but they are a much different exercise for a six-foot-ten, 340-plus-pound offensive lineman like Swoboda. He had always dreamed of playing Division I college football, and he was determined to impress the coaches, so he tried his hardest. In the evenings, he guzzled Gatorade in his room and prayed that he could keep going. "I wanted to make a name for myself, trying to push myself as hard as I could," he says.

He pushed himself all the way until the third afternoon, when he finished a sled drill and promptly blacked out.

He woke up in a hospital room with a fuzzy male figure. As soon as he opened his eyes, the fuzzy figure rushed over and gave Swoboda a huge hug in the bed, sobbing, "I love you, man."

Swoboda saw that the man was wearing a blue UVA polo and replied, "I love you too, Coach."

It was his dad. Swoboda had been in a coma for three days.

Doctors explained that he had experienced a severe heatstroke, meaning his body temperature had risen to a point where he could not cool back down. Typically, someone is at risk of exertional (exercise-induced) or passive heatstroke* when their core temperature rises past about 104 or 105 degrees F (40 degrees C), although it varies from person to person. Swoboda's core temperature had hit 109 degrees F.

He had clung to life by a thread. It didn't matter that he was an eighteen-year-old athlete in his physical prime. Heatstroke can strike down anyone, as famed American physician James J. Levick observed in 1859: "It strikes down its victim with his full armor on. Youth, health, and strength oppose no obstacle to its power; nay, it would seem . . . to seek out such as these."[4]

The only reason Swoboda even had a chance was because he had been at least partially cooled down by trainers who rolled him onto a blue plastic tarp and dumped buckets of ice over his body before rolling him back and forth, a procedure known as "TACO" (for "Tarp-Assisted Cooling with Oscillation"), before the ambulance arrived. That had likely saved his life. But his football career was clearly over, due to his extensive internal injuries. The University of Virginia offered him a medical retirement, meaning they would still honor his scholarship but he would not need to play.

No thanks, Swoboda said. He wanted to play again. But his coaches were skeptical, and the university would not allow it. His dad found out about KSI, one of the few places in the world that specializes in rehabilitating victims of severe heatstroke. So he began working with Huggins and Douglas Casa, the director of KSI.

The first step was heat tolerance testing, which Swoboda flunked, just like me. The test also revealed he sweated out a staggering amount of fluid, on the order of four liters (more than a gallon) of liquid per

---

* Exertional heatstroke is different from classical heatstroke because the former tends to happen suddenly, while the latter, also known as "passive" heatstroke, develops over hours or even days, typically in hot, humid conditions, and tends to affect primarily the elderly and people with preexisting illnesses.

hour as he worked in the heat. A normal adult male will sweat out about one liter per hour. All this would have been good to know *before* his near-fatal heatstroke, but better late than never.

Huggins and his KSI colleagues designed a rehabilitation program to try to bring Swoboda back to the point where he could play again. The "medicine" they prescribed was the exact same thing that had almost killed him: exercising in the heat. Hard. On purpose.

Every day, back in Virginia, Swoboda would wake up at 4:30 a.m. and swallow a temperature-sensing capsule the size of a horse pill. Then he would wait until the hottest part of the day to go work out—running sprints and doing more up-downs, outside in the full Southern sun. "I'd wake up every morning and go, 'I have to go give myself a heatstroke again,'" Swoboda told me.

To prevent that from happening, his trainer watched him every minute, tracking his temperature via his phone as the capsule made its way through Swoboda's digestive tract, transmitting temperature readings all the way. The goal was to work hard enough that his body temperature rose to a certain level, between 101 and 102 degrees F, but not so high that he risked another heat illness.

After three weeks of these workouts, and untold gallons of sweat, he went back to KSI for another heat tolerance test—and flunked. He did another three-week bloc of heat training and flunked the test again. And again. He finally passed on his fourth try, which meant he could be cleared to play, a year after he had nearly died on the practice field.

"I broke down crying in the locker room when I finally passed the test," he says. "I never thought I would play football again."

Through heat training, Swoboda became acclimated to the heat in a way that he never had been in his life, despite having grown up in Florida. His physiology changed; his body became much better at maintaining homeostasis—which he says is his favorite new word that he learned in college—even in extreme heat and humidity. Homeostasis saved his life.

More surprisingly, he found that his ordeal had made him a better player. He had been a middling recruit, likely destined to ride the bench. When he came back, he earned a starting spot as an offensive tackle, playing three seasons for UVA and then a fourth for the University of Central Florida, close to where he had grown up. After graduating, he performed well enough at the NFL Combine that he qualified for the 2023 NFL draft.*

Heat had nearly killed Ryan Swoboda, but it had also made him stronger, fitter, and more resilient, both physically and mentally. Heat had been the cause of his problems, but also the solution. Ultimately, it had healed him, protected him from future harm, and made him a better athlete.

I was hoping that it might do the same for me.

Minus the almost-dying part.

---

* Swoboda played on practice squads for a handful of teams before being medically retired in 2024. He now works in real estate in Florida.

2

SWEAT

No one has ever drowned in sweat.
—LOU HOLTZ

On a bitter cold January afternoon in 1774, a group of learned gentlemen gathered in a London town house to conduct a strange and highly dangerous experiment. The question at hand: How much heat can the human body tolerate?

Their ringleader was a prominent Scottish-born physician named Dr. George Fordyce, a member of the Royal College of Physicians who had earned a reputation for eccentricity. Every afternoon, he would dine at an upper-crust establishment called Dolly's Chop House, usually consuming a pound and a half of beef rump steak, plus a chicken breast or some fish, washed down by a tankard of strong ale, a bottle of port, *and* a beaker of brandy. Thus fortified, he would go and lecture to his private medical students, which must have been quite entertaining.

Fordyce was fascinated by body temperature; he was hard at work on a groundbreaking multivolume treatise on fever. But he was also curious about how external heat might affect our physiology. In the rented town house, his servants had rigged up a system of stoves and flues to heat three rooms to extreme temperatures, making it a kind of predecessor of the KSI heat lab. When the rooms were ready, the experiment began: Fordyce stripped down to just his shirt and walked into the first room, which was warmed to 90 degrees Fahrenheit—like the warmest English summer day. Within five minutes, he began to "sweat gently," according to a colleague, Charles Blagden, who was recording the experiment. Satisfied, the pantless doctor moved on to the next room, which was warmer, at 110 degrees F. Here, he lasted only half a minute before he began sweating profusely. Before long, Blagden observed, "his shirt became so wet he was obliged to throw it aside, and then the water poured down in streams over his whole body."[1]

Now naked and drenched in sweat, the doctor moved into the third, even hotter room, at 120 degrees F. Soon his entire body flushed red and his heart raced, reaching 145 beats per minute, which was alarming given his gluttonous diet. But even after twenty minutes in this impromptu sauna, his body temperature had hardly budged, according to a thermometer placed in his mouth. His colleagues also measured the temperature of his urine, hopefully with a different thermometer, and found to their astonishment that it too had remained at just 98 to 99 degrees F. How was this possible?

The gentleman scientists were trying to answer a question that had been posed to the Royal Society years earlier by one Henry Ellis, who had served as colonial governor of Georgia during the 1750s. While walking around Savannah on a fiercely hot summer day, Ellis was stunned to see that his personal thermometer—dangling from the brim of his hat—said that it was 105 degrees F outside, a staggering level of heat.* Yet when he pressed the bulb of the thermometer against his skin, the highest it ever read was 97 degrees Fahrenheit. Why wasn't his body just as warm as the air? More to the point, he wondered in his letter, how was he even still alive, in such "prodigious" heat?

The local Muscogee Creek Native Americans, with whom Governor Ellis had been negotiating a treaty, had been living in that very same prodigious heat for hundreds of years—trading, traveling, farming, even building large pyramid-like earthen temples and burial mounds. But for Ellis and his fellow British-born colonists, this was uncharted territory; the Georgia summers were far hotter, and the winters colder,

---

* Assuming that Mr. Ellis's thermometer was accurate, 105 degrees F would have equaled the highest temperature ever recorded at Savannah.

than anything they had experienced at home. Eventually, the heat took such a toll on Ellis's health that he was forced to leave his post.

In his letter, he moaned, "What havoc must this make with a European constitution?"

A few days later, Fordyce and Blagden returned to the rented townhouse to push the limits even further. This time they had three other gentlemen in tow, including the celebrated naturalist Sir Joseph Banks, who had accompanied James Cook on his first voyage of discovery around the world in the 1760s and had personal experience with steamy climates. Fordyce had had one room heated all the way up to 150 degrees Fahrenheit by a woodstove that glowed cherry-red. All of the gentlemen entered the room—keeping their clothes on this time—and while they quickly became very hot and sweaty, once again their body temperatures did not rise.

Mystified, they went to dinner while the stove kept burning. By the time they returned, presumably well stuffed and slightly tipsy, the metal objects in the room had become too hot to touch, and even some of the thermometers had shattered. Bravely, they went back inside anyway, bringing with them a hunk of steak and a basket of eggs. The eggs were soon roasted hard, and the steak cooked tougher than a shoe, but the men survived, albeit with shaking hands and ringing ears—rather like me when I tried to hang with the Blister Boys. Yet when Blagden pressed the thermometer to his own skin, his body temperature stubbornly remained at 98 degrees F.

It quickly devolved into a contest to see who could withstand the most heat, as the room got still hotter. Sir Joseph Banks, veteran of the tropics, eventually won, tapping out at an incredible 211 degrees Fahrenheit. He had sweated right through his clothes but was otherwise fine. Then they all went outside, not bothered at all by the frigid January air.

This was not merely an idle gentlemen's exercise. At the time, the British were busy colonizing India, North America, parts of Africa, and the Caribbean, which meant that her sailors, soldiers, bureaucrats, and grandees were getting a taste of steamy Equatorial climes—a drastic change from chilly, rainy England, where a "warm" summer day might top out in the mid-seventies Fahrenheit. British colonists like Governor

Ellis suffered mightily in their sweltering colonial outposts. They simply weren't accustomed to the heat.²

Other European powers were also expanding their influence in places with very hot climates, like South America and Africa. This exposed the newcomers to novel and deadly diseases, like malaria and yellow fever, that felled them like flies. The question of how to cope with extreme heat was therefore not only a scientific question, but a national security issue, not to mention a marketing challenge, as governments sought to persuade would-be settlers that their prospective new homes were, in fact, inhabitable.

These places were, of course, already inhabited—by Native Americans, Africans, Indians, and others, who appeared to have no difficulties surviving in the heat. According to the "science" of the eighteenth and nineteenth centuries, however, darker-skinned people possessed a natural affinity for hot weather that white people allegedly lacked. A paper from 1898 titled "Acclimatization of Europeans in Tropical Lands" discussed the long-running controversy over whether Caucasians could survive more than a few generations in the tropics. The author declared: "Crossing with native stock or with immigrants better adapted to the new environment is considered by many as the best and most rapid way of securing acclimatization. Intermarriage is said to be the [key to the] success of Spaniards and Portuguese in Mexico and the Philippines."³

While racial theories about heat tolerance persisted, geographical exploration and colonization had prompted many new discoveries in medicine and biology, while spurring technological advancement. The shiny new Fahrenheit thermometer that Governor Ellis sported on his hat represented both the height of fashion *and* the cutting edge of eighteenth-century technology.

Scientists had been struggling to figure out how to measure temperature since ancient times, mostly without success; the problem had even stymied such brilliant Renaissance and Enlightenment minds as Galileo and Sir Isaac Newton.* It wasn't until 1714 that Daniel Gabriel

---

* Newton was also the first to try to quantify temperature objectively, scribbling a scale of "degrees" on the shaft of his crude thermometer in 1701. Notable reference points included the temperature of melting snow (zero degrees); "the heat at midday about the month of July" (6 degrees); body temperature (12 degrees); and the point at which water "boils vehemently" (34 degrees).

Fahrenheit invented and perfected his familiar closed-bulb mercury thermometers, which were by far the most accurate such gauges that had yet been made—and which remain in use today. Only a few small, backward nations still use his Fahrenheit temperature scale, which has been supplanted by one attributed to his Swedish competitor, Anders Celsius.*

Fahrenheit's newfangled device had helped these gentleman-scientists, Blagden and Fordyce, to uncover a key insight into human biology, one that is fundamental to our history as a species. Based on their experiment, they reasoned that the human body must have some special ability for enduring or resisting heat—and that this power of resistance might be improved. In other words, the colonists could learn to adapt to their new homes.

As Blagden wrote, presciently: "Probably both the power of destroying heat, and the time for which it can be exerted, may be increased, like most other faculties of the body, by frequent exercise."[4]

The word he was searching for, to describe the "power of destroying heat," is one that you will be hearing often in the pages to come: *thermoregulation*.

All animals thermoregulate—control their body temperature—to some extent. Some are better at it than others. Cold-blooded creatures such as snakes and lizards are known as ectotherms, meaning their temperature fluctuates with that of their environment; they produce little to no body heat of their own. This is why rattlesnakes sometimes slither out onto a warm stone patio as the sun goes down, scaring the daylights out of homeowners. Warm-blooded mammals and birds are endotherms, meaning they generate heat within their bodies—typically via metabolism and the action of their muscles—and retain that heat thanks to the insulation of fur, fat, and feathers.

Heat is essential to life, but so is thermoregulation. Most mammals and birds require a steady internal body temperature within a narrow range: not too hot, not too cold, but just right, like Goldilocks. In order to maintain

---

* Although much of the world now measures temperature using the Celsius scale, the original version of that scale ran backward—that is, water boiled at 0 degrees C and froze at +100 degrees C. It was reversed by others after his death.

homeostasis, Ryan Swoboda's favorite word, they must balance the heat they produce with the heat they dissipate into the environment, striving always to remain within that Goldilocks zone. And while the idea of thermoregulation is simple, the mechanisms by which it is achieved are quite complex, involving our skin, nerves, muscles, blood vessels, heart, and brain.*

Let's go back to our car analogy. If we are driving up a long mountain pass, our engine will generate an enormous amount of excess heat. Unless there is some way to dissipate that heat, the engine will quickly overheat, especially on a hot day. Which is why automobile engines have pumps and hoses and radiators and fans and coolant—all to keep the engine at an optimal temperature.

Mammals are somewhat similar: most function best at a body temperature of around 98 degrees F (37 degrees C), give or take. But unlike a car, most mammals do *not* have much wiggle room. Our optimal body temperature is already quite close to the upper limit that living beings can tolerate—the *critical temperature limit*, in physiology-speak (or, for our metaphorical car, the red line). Going just a few degrees beyond that limit—above about 104 or 105 degrees F—can put us at risk for serious illness or even death.

But these limits are not absolute. Professional cyclists and runners have finished races with internal body temperatures as high as 107 degrees F (41.5 degrees C) or higher, although they were likely only that hot for a relatively brief period of time.[5] In one famous experiment from the 1970s, a then–PhD student at the University of California, Santa Barbara, named Michael Maron ran the Santa Barbara marathon with a friend, both with rectal probes inserted and a team in a follow car taking temperature readings every mile. Despite the probes, both runners finished the race in well under three hours, one of them (Maron) with a rectal temperature above 107 degrees F—yet with no sign of heat illness and no apparent decline in running performance.[6]

---

* The best technical definition of thermoregulation that I have found comes from a review by the eminent physiologist Elizabeth Repasky: "Heat is the end product of nearly all of the energy released in the body, and its production is essential for all life. To maintain a very narrow range of core body temperature, the heat produced from metabolic reactions must be constantly balanced with that absorbed from or dissipated to the external environment; this thermal homeostasis is maintained through extensive neural, vascular, and biochemical mechanisms collectively known as thermoregulation." In other words: It's surprisingly complicated.

Above that, however, it gets dicey. Ryan Swoboda nearly died because his core temperature hit 109 degrees F. Other athletes have survived still higher core temperatures, as high as 110 degrees F or even 113 degrees F, provided they were cooled quickly (best method: full-body immersion in a tub of ice water). The most extreme case I was able to find involved an extremely lucky fifty-two-year-old man who managed to survive a core body temperature measured at 46.5 degrees C (115.7 degrees F) during an Atlanta heat wave in 1980. (He was already in a coma when he was brought to the ER, thanks in part to the quart or two of vodka he had drunk that day while cooking over a hot stove. Hospital personnel cooled him aggressively using bags of ice.*)

Such cases test the limits of biological tolerance. Above about 114 degrees F, things literally start to fall apart. As key cellular proteins "denature," or lose their structure and hence their function, our cells and certain organs simply stop working properly. Cell membranes break down, and the intestinal lining may dissolve, leaking waste products and toxins into the bloodstream, which is as awful as it sounds. This, in turn, ignites a massive immune-system reaction that can ultimately cause our vital organs to fail, leading to a horrible and agonizing death.

To avoid this, evolution has devised some clever strategies for helping mammals keep cool. The enormous, floppy ears of elephants, for example, are ideal for radiating heat back into their environment. The tails of certain rodents serve a similar function—as do our own hands and feet. Kangaroo rats and other desert rodents help stay cool and hydrated by extracting every bit of water from their feces before they poop. Many other mammals simply stay in shade or shelter during the heat of the day, coming out to forage and hunt only at night. This is *behavioral* thermoregulation, and humans are quite good at it too. When we are hot, we move into the shade or find a cave or a body of water in which to bathe or swim; also, we can remove our clothing, layer by layer.**

---

* Another, even stranger, case involved a man who got into a fender-bender with a police car on a hot day, swallowed the methamphetamine he was carrying, and then sprinted away from the cops. His core temp topped out at 113 degrees F, yet he somehow survived.
** Sometimes, when riding my bike on a hot summer day, I will take off my shirt, which also counts as thermoregulation, and feels wonderful.

(The ideal environmental temperature for a naked human, in case you were wondering, is in the mid-80s F, or around 30 degrees C.)[7]

The absolute master of hot-climate survival is, of course, the camel. Its signature feature, its hump, serves to store food in the form of fat (and not water, as many people believe)—but more importantly, it shades and insulates the camel's body from the blazing desert sun. The fur on top of a camel's hump can get as hot as 175 degrees F, while its internal body temperature stays at a comfortable 100 degrees F. (Although as the desert climate swings from extreme heat at midday to severe cold at night, the camel's body temperature follows along, climbing as high as 113 degrees F during the day and dipping as low as 93 degrees F on cold desert nights—extremes that would quickly incapacitate or kill nearly any other animal.)

Not all creatures are so well adapted, but all animals thermoregulate in some way. My dog, Bode, for example, has his own strategies for thermoregulation. When he is not foraging for dropped donuts and half-eaten pizza slices, he loves nothing more than feeling the warmth of the sun in his old hound-dog bones as he basks in our backyard. As the day heats up, he will reluctantly get up and move into the shade and then, eventually, maybe dig a small hole in the garden to uncover some nice, cool dirt and lie down in that for a while, much to the annoyance of his mom. Finally, well past the point at which I begin to worry about him getting "too hot," he will yowl at the back door to be let into the house.

All of this—the shade-seeking and the water-drinking and the eventual coming inside—counts as thermoregulation. But one thing that Bode absolutely will *not* do on a hot day like I have just described is get up and run anywhere. Or even walk, really; a trip around the block is almost more than he can endure. I, on the other hand, might hop on my bike, or walk to the coffeeshop, or do some gardening, and feel fine—because only one of us can sweat, and it isn't him.

Sweat glands come in two basic types. All mammals, including Bode the dog, possess the first kind, which are called *apocrine* sweat glands. Always located at the base of a hair follicle, apocrine glands produce a viscous, cloudy liquid that contains not only water but also some fats, proteins, and miscellaneous hormones and other things, including ammonia. Ooey and a little gooey, this is the sweat that gives sweat a bad name, because it feeds all manner of skin bacteria that, in turn, emit the noxious smells that we recognize as "body odor" and that the bacteria call "our farts."

"That's the gross stuff," confirms Andrew Best, a professor of evolutionary biology at the Massachusetts College of Liberal Arts. Other examples of apocrine-gland secretions include ear wax, the fat droplets in breast milk, and the smelly spray of the skunk.

The other kind of sweat glands are *eccrine* sweat glands. These are much less gross, emitting only water and salt and other electrolytes. Mammals such as mice, monkeys, and dogs like Bode typically have some of these eccrine glands as well, mostly concentrated around their feet or paws. Humans, on the other hand, have eccrine sweat glands all over our bodies—millions of them, not only on our hands and feet but on our chest, back, arms, forehead, and just about everywhere else. We are, hands down, the sweatiest animal on the planet. And our eccrine sweat glands serve a novel purpose: cooling.[8]

The eccrine sweat gland is a marvel of natural engineering, elegant in its simplicity yet powerful in its function. It consists of a small, coiled tube just beneath the surface of the skin that resembles "a ball of yarn that a cat might play with," says David Martin, one of the KSI lab assistants (now director of science and research at Perry Weather, as well as a cycling performance coach). This "secretory coil" (as it's called) leads to a tiny duct that travels through the epidermal layer to a pore. The whole thing is minuscule: The coil is only about half a millimeter in diameter, and the duct can be three millimeters long or so. Yet this tiny organ was crucial to our evolutionary success.

Thermoregulation is a highly complex process, as our body seeks to balance the heat it generates within with the heat it dispels into (or absorbs from) the external environment. Sweating is actually *not* our first line of thermal defense; water is so precious, so essential to life, that it would be foolish to waste it unnecessarily. Sweating is the crucial last step in a chain of events that most of us aren't even aware of.

Let's say you go for a walk on a hot summer day. At first, you feel fine, neither hot nor cold ("thermoneutral," in science-speak). But as you get warmer, from the effort of walking, the warmth of the air, and perhaps the direct heat of the sun, you will unconsciously begin to divert more blood to your skin and your extremities, especially the fingers and toes. Because your skin temperature is several degrees cooler than your core temperature, typically around 93 degrees F versus ~98.6 degrees F for your core, heat will travel from inside your body to the skin and then into the environment—because, as Sir Isaac Newton first observed, heat always flows from warm to cold. (This is also why days when the air temperature is warmer than 93 degrees F can feel so uncomfortably hot.)

If it's really hot out, your skin blood flow might increase by as much as sixteenfold, meaning that effectively your entire blood volume is whooshing just beneath the surface of your skin each minute, especially to our hands and feet, our natural "radiators." At this point, your skin might start to turn a little pink, even red. Cooling is so important that your body will divert blood flow from your digestive system, your muscles, your vital organs, and even your brain to try to maintain normal body temperature.*

Because it is so important to our health and survival, our body temperature is very closely monitored—far more so than the temperature in our home. While we might have only one or two thermostats in our home, our bodies have millions of temperature-sensing channels scattered all over our skin surface, in our gut, in our key organs, in our muscles, and in our brain. These all work together to provide input about our thermal status to the hypothalamus, which is responsible for regulating our body temperature.

The interplay among all these "thermostats," known as temperature receptor pathways or TRPs, is highly complex and not fully understood. Experiments in chimpanzees (who also have eccrine sweat glands, like us) have found that warming up their brains using surgically implanted

---

* Most people think our core temperature is *always* 98.6 degrees F—but in fact it varies from person to person, and hour to hour. And many people are below 98.6 degrees F. One large study of more than six hundred thousand patients' medical records found that their average body temperature reading was only 98.0 degrees F (36.7 degrees C). Also, our core body temperature typically rises during the day, when we are active, and then drops during the night. But it can also change due to other factors, such as our metabolic rate, our level of activity, whether we have an infection (i.e., fever) or are experiencing symptoms of menopause, and even (as we will see later) our mental health status.

heating elements will cause them to start sweating, which makes sense. But if researchers cool the animals' skin while still heating their brains, they will not sweat.

Making it even more complicated is the fact that there are multiple types of temperature-sensing channels, some attuned to heat and others to cold, which is equally if not more dangerous to you as a human. Also, different channels are sensitive to very specific temperature ranges, and some can also detect sensations such as wetness, salinity, or even the presence of capsaicin, the chemical that makes hot peppers taste spicy.*

Still others are attuned to extreme heat, to protect us from injury from hot water or hot objects, like the handle of a metal pan. The important point is—for our purposes—that your body monitors its internal and external temperature(s) *extremely* thoroughly and closely.[9]

Eventually, as you continue to walk on that hot day, or perhaps even begin to jog, enough heat will accumulate in your body that your temperature will increase. At some point, your brain will decide that it's time to start sweating. Now your eccrine sweat glands are summoned to action, by a signal from the sympathetic (fight-or-flight) nervous system.** This signal kicks off a series of biochemical reactions that force potassium and sodium from neighboring cells into the "yarn ball" coil. Water will then follow, coming from those neighboring cells and the intercellular space into the sweat gland duct, and then making its way to the surface of our skin.

As this happens in hundreds of thousands or even millions of sweat glands all over the surface of our body, all at once, we notice a sudden dampness, that familiar bloom of moisture across our skin that appears when we begin to dance, play games, or make love. As it goes on, we can become downright soaked, shiny, and slick. Now we are sweating.

---

* David Julius won the Nobel Prize in Physiology or Medicine in 2021 for his work on how the body senses heat, cold, and spicy peppers.

** This is also why we often break into a sweat in stressful situations, such as a job interview or a first date, where our fight-or-flight system is activated. One evolutionary explanation for this might be that in case we actually do end up needing to flee, our flop-sweat acts as a form of precooling. Also, there is some evidence that eccrine sweat glands help with wound healing, which also makes sense in this hypothetical fight-or-flight scenario.

How Sweating Works

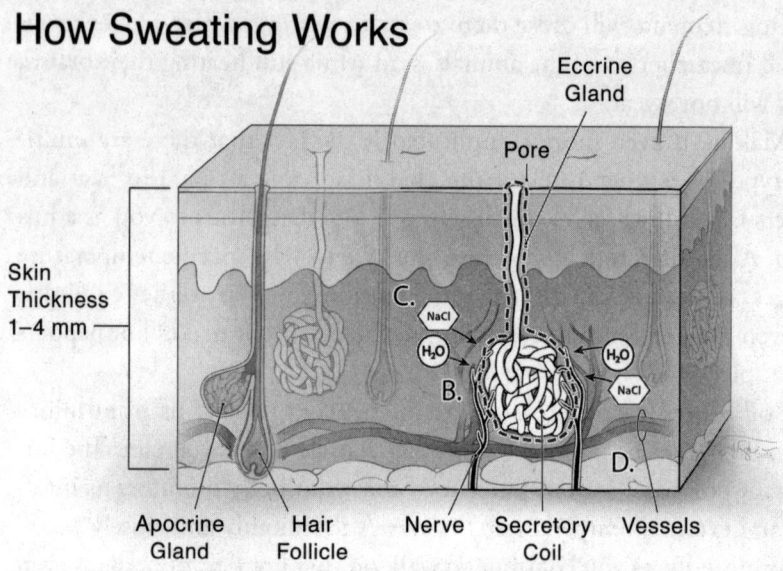

Evaporative Process

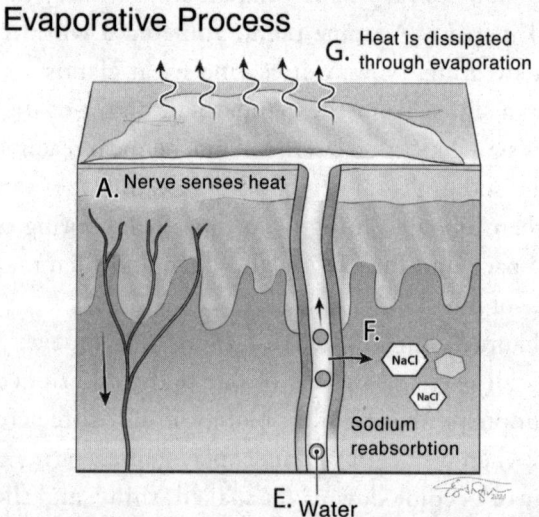

Cross-section diagram of apocrine (top L) and eccrine (top R) sweat glands. When skin nerves (A) sense heat, the brain activates eccrine sweat glands in the skin (B). Specialized cells around the secretory coil expel salt (NaCl), and then water, into the sweat-gland duct (C); meanwhile, blood flow in nearby vessels (D) increases dramatically. As the liquid (E) makes its way to the skin, some of that salt is reabsorbed by other specialized cells (F) before the water reaches the skin surface (G), where it evaporates and dissipates heat, cooling the blood and the skin. (Illustration by Eo Trueblood)

But the sweat that puddles on the ground or soaks into our clothing does almost nothing to cool us; it is essentially wasted. "Dripping sweat does not help you," says Christopher Minson, a scientist at the University of Oregon who has studied sweat and heat adaptation for two decades. "It *has* to evaporate."

This is the key step, the magic: evaporation. Sweating is not merely about removing warm water from the body. The transition of sweat from liquid into vapor is what ultimately cools us. "That phase shift takes away energy, which is the heat you produced," Minson explains. The evaporating sweat cools the skin and the blood flowing directly beneath it, which then returns to the heart to help cool the body's core.

It does so extremely effectively: Each milliliter of sweat that evaporates from your skin (basically, a few drops) carries away 580 calories' worth of heat energy. That adds up. If you sweat out a liter of liquid in an hour—a fairly typical sweat rate for an adult male exercising in the heat—and all that water completely evaporates, it carries away enough thermal energy to cool your entire body by about 10 degrees F (assuming you weigh 175 pounds).* That's a lot of heat energy being carried away by your sweat. The caveat is that not all of our sweat evaporates, especially in a humid environment such as a hot yoga class or a summer afternoon in Miami or Mumbai, which is one reason why such environments feel so much hotter and more uncomfortable than hot and dry places.

In both dry and humid situations, however, our eccrine sweat glands give us a lethal advantage over all other mammals. If my dog, Bode, tries to run with me on a warm day, for instance, he will try to cool himself by panting, with his tongue flapping in the breeze. The water evaporating from his tongue will also cool him off, a little. But even if he has the longest, floppiest, cutest pink tongue of any dog ever, Bode can shed only a relatively small amount of heat in this way. Fairly soon, he will need to slow down or stop. I can keep on going, on the other

---

* Another way to visualize this: Evaporating a liter of sweat represents the same loss of thermal energy as cooling seven liters (approximately two gallons) of water from boiling down to room temperature. What I'm trying to say is, the evaporation of sweat removes a tremendous amount of heat from the system.

hand, because I have many, many more eccrine sweat glands than Bode or any other mammal does, spread out over my entire body surface.[10]

"We've basically turned our entire body into a giant, wet tongue," says Daniel Lieberman, who has studied sweat, evolution, and endurance sports for decades and has a gift for memorable images. "We can really lose an enormous amount of heat this way, so we can run farther and run hotter."

All thanks to a tiny, almost invisible little structure embedded in our skin.

Eccrine sweat glands likely first appeared in mammals about forty million years ago, according to Best, but it was only two to four million years ago that sweating began to emerge as a cooling strategy in our primate ancestors.[11] "It's sort of a mystery why eccrine glands evolved in primates initially," he says. "I think it's kind of an accident of evolution. But once they did, they were incredibly useful."

Yana Kamberov began studying the evolution of sweating as a graduate student in Lieberman's lab. Now a geneticist at the University of Pennsylvania, she has spent much of her career trying to understand how and when we acquired this talent. "I love sweat," she says. "And it struck me that the skin organ most important to human survival is one of the ones that we know the least about."

Sweat glands themselves are difficult to study; they can't be isolated in a petri dish, and harvesting live sweat glands requires persuading volunteers to undergo a painful procedure called a "punch biopsy." There aren't many takers. So we don't fully understand how sweat glands operate, why some are activated while others remain dormant, and why some individuals sweat so much more than others. And because sweat glands do not fossilize, we have almost no idea when these little mini-organs first appeared in our prehuman ancestors. To understand how—and, as important, *when*—we acquired this superpower of sweating, we must look to genetics.

Kamberov began by comparing humans and our closest primate relatives, chimpanzees. Evolutionary history suggests that humans and chimpanzees shared a common ancestor about five to seven million years ago, before we split into two branches of the evolutionary tree.

There are many important differences between us, obviously, but one of the most obvious is the fact that humans are far sweatier than chimpanzees, with many more eccrine sweat glands—ten times as many—and much less visible body hair.[12]

Kamberov set out to try to understand what had happened genetically on the evolutionary road from early primate to human. Long story short, she found the key to the evolution of human sweating in an unlikely place: toe sweat. Specifically, toe sweat in mice.

Evolution doesn't typically create things from scratch; rather, it modifies and repurposes and builds on existing structures. Gills become ears, a snout becomes a nose, fins become legs, a finger becomes a thumb, and so on, as changes are filtered by natural selection. A small brain becomes a bigger and more complex brain. As noted earlier, mice do have some eccrine sweat glands, mostly concentrated around their paws. At some point on the long road to *Homo sapiens*, Kamberov theorized, evolution had "decided" to move eccrine sweat glands from some mammals' paws and feet to the entire rest of their bodies, and to use them for a completely novel purpose: cooling.[13]

To demonstrate this, she genetically modified some laboratory mice with a specific "transcription factor"—basically a boss gene that orchestrates the work of other genes*—called Engrailed-1 or *EN1*. While we are developing in the womb, Engrailed-1 is deeply involved in the process of creating skin, mammary glands, hair, and teeth, among other essential features. It also helps to create sweat glands.[14] It's a busy little boss gene. By turning up the activity of Engrailed-1, along with another orchestrating gene called an "enhancer,"[15] Kamberov found that these genetically modified mice developed many, many more eccrine sweat glands in the soles of their little mouse feet.

---

* Most people know that genes code for proteins, one to one: You punch in the DNA code, and out pops a specific protein, like a soda or candy bar from a vending machine. In fact, the process is far more complex, more like assembling a Chipotle burrito. Specific ingredients (the proteins) must be added in the correct amounts, in the correct sequence, with the correct techniques, or the whole thing ends up a mess. Someone needs to orchestrate this process—and in genetics, this is the job of transcription factors. They are the burrito-makers of biology.

"They were not very happy mice," she says of the sweaty-footed rodents. But they had helped to prove her point.

Next, she looked more closely at the enhancer and found that it had been mutated no fewer than ten times in the prehuman genetic history, within a relatively short period—in ways not found in any other primates.* Each successive mutation created still more eccrine sweat glands. "It's been targeted over and over and over again by evolution," she says. "Natural selection just went crazy."

While this was happening, our primate ancestors gradually lost their coarse body hair as well. This enabled sweat to spread across their skin and evaporate more easily—although that body hair did not exactly disappear, Kamberov notes. Rather, the hair follicles simply shrank until they were nearly invisible. "We are essentially as hairy as a chimp," she says, "but they are fine vellus hairs."**

We have retained thick, dense hair in certain obvious places—armpits, crotches, atop our heads, where it helps protect us from the sun. That too was a thermoregulatory adaptation that took place thanks to natural selection. But as they lost most of the rest of their body hair, our primate ancestors gained millions of new eccrine sweat glands, which proliferated wildly across their entire bodies—everywhere but on the lips and the head of the penis.

A typical adult human has between two and four million eccrine sweat glands, a staggering number, but they are not equally distributed. We tend to have more sweat glands per square inch on our chest than on our back, and more on our forehead than on our arms. The sweat of one's brow is meant to help keep the brain cool. But we *still* have the greatest concentration of eccrine sweat glands on our hands and feet—just like our mammalian ancestors (and Bode the dog). Something to keep in mind when you are tempted to go sockless on a plane.

All these mutations were already in place some eight hundred thousand years ago, says Kamberov. Other humanlike species, including the

---

* Except for a small species of monkey called patas monkeys.
** The same thing happens to men who appear to lose their hair in middle age. They (um, we) are not technically "bald," because the hair follicles are technically still present. Unfortunately, however, the actual hairs have shrunk down to near invisibility. ☹

Neanderthals and Denisovans, share the same basic set of sweat-related genes. She believes that natural selection had already been working on the sweat problem for a long time by then, although it is not yet clear when this evolutionary march toward sweatiness began. Regardless, the fact that this trait already existed in our ancestors eight hundred thousand years ago suggests that sweating, and being able to thrive in very hot conditions, played a major role in the rise and eventual dominance of the human species.

"Sweat is fantastic, and humans are the best sweating animals on the planet," says Chris Minson. "There is no animal that can beat us at sweating."

Many animals, however, can beat us quite handily at other things, notably sprinting, clawing, biting, killing, and eating. As our primate ancestors acquired the ability to walk upright more than four million years ago, they sacrificed much of the speed and agility they had previously enjoyed. With only two legs to propel them rather than four, they became slow and thus easy prey for lions, hyenas, saber-toothed tigers, and whatever else might like to snack on fresh, easily caught *Australopithecus* meat.

Sweating helped us escape that evolutionary dead end, Daniel Lieberman explains. As they acquired the ability to cool themselves effectively, these prehumans were now able to forage for food and water during the heat of the day, when most of their fur-covered, non-sweating predators and competitors were hiding in the shade.

"There was strong selective pressure to forage in the middle of the day," says Lieberman. "That would have led to strong selection for characteristics that make us good at dumping heat—miniaturization of our hair and proliferation of our sweat glands."

Even better, this newfound talent for sweating helped humans graduate from prey to predator, because it enabled our ancestors to run long distances in hot weather. Because they could sweat, and thus stay cooler, humans and humanlike species such as Neanderthals were able to chase their prey for long periods of time, as a University of Utah biology professor named David Carrier theorized in the 1980s.

According to Carrier's "persistence hunting" hypothesis, first proposed in 1984—and later revived, refined, and built upon by Lieberman and others[16] (and popularized by my friend Christopher McDougall in his bestselling book *Born to Run*)—our ancestors would literally run game animals to death, chasing them until they died. This seems impossible; who can outrun a deer? But while deer and antelope and other large four-legged animals can easily outrun humans over short distances, they cannot sustain this speed for a very long time, especially in warm weather, because they get too hot. Eventually, these large fur-covered mammals will have to stop and rest (and pant)—allowing us, or our ancestors, to catch up to and eventually kill the exhausted prey.

Few of these hunts have ever been witnessed or recorded, so it remains a hypothesis, but they likely would have required those ancient hunters to walk or slowly run after their prey, for periods ranging from a few hours up to a couple of days, perhaps with a final burst of speed to finish the kill.* Such hunting tactics were not limited to hot climates either: Ten-thousand-year-old cave paintings in the Altai Mountains of Siberia and western China depict ancient hunters chasing down their prey on skis.[17]

All of this happened thanks to our ability to sweat, which was like evolution's nuclear weapon, giving us a devastating advantage over our prey. Our power of thermoregulation meant that we—and our ancestors—were able to operate at higher levels of exertion in hot conditions, and sustain a higher body temperature, than nearly all other mammals. Sweating lets us push the limits of our heat tolerance, the upper critical temperature I mentioned earlier.

"Once you get above that, you're basically dead," Lieberman explains. "But because we're good at sweating, we're better at being active at temperatures closer to our upper critical temperature, because we can

---

* One of the few eyewitness accounts of such a pursuit is captured in the 2000 documentary *The Great Dance: A Hunter's Story* by Craig Foster, who also directed *My Octopus Teacher*. Foster and his crew followed a group of tribesmen in the Kalahari Desert of Botswana as they ran down and killed a kudu in a ceremonial hunt. Hard to find at this writing, but I accessed it here: https://archive.org/details/TheGreatDanceAHuntersStory2000.

dump heat better than any other mammal. People can run marathons in ninety-degree heat."

Or ride a bike one hundred miles in 100 degrees F heat, if that's your jam. The point is, we are designed to push our limits in the heat. And while bad things can happen if we push it too far, some good things also happen to us when we sweat, as I was soon to learn.

3

GOOD HEAT

The sauna is a poor man's pharmacy.
—FINNISH PROVERB

A FEW days before my encounter with the Blister Boys, I rode the gleaming Helsinki Metro four stops out from the city center and emerged in suburbia—which, this being Finland, turned out to be a lovely island with parks and beaches. Outside the station, I spotted an older man carrying a nylon shopping bag who seemed to know just where he was going. I could guess where he was going too: the Finnish Sauna Society, the *Suomalainen Saunaseura*, a place that is as revered among sauna lovers as the Old Course at St Andrews is among golfers.

Built in the runup to the 1952 Helsinki Summer Olympic Games, the *Saunaseura* was intended to provide a recovery spot for the Finnish athletes and to showcase Finnish sauna to the world. It sits on a wooded, rocky peninsula, with three log-walled saunas attached to a main clubhouse. Risto Elomaa, the president of the Finnish Sauna Society, was waiting for me in the parking lot. Tall, lean, and still sparkling with energy in his late seventies, he comes here four times a week and spends most of the rest of his time spreading the sauna gospel. First lesson: The Finns pronounce it "SOW-nuh" (as in a female pig), not "SAW-nuh" (as in a tool for cutting wood).

He ushered me through check-in, where I was issued a scratchy white towel for drying and a smaller black towel for sitting on.

Inside, not much seemed to have changed since 1952. Women bathe on Mondays, Thursdays, and Saturday mornings. Men have the rest of the week. There is a modest lunchroom serving a famous salmon-and-potato-dill chowder. It reminded me of some ultra-exclusive, shabby-genteel golf club, except that everyone was naked; bathing suits are tolerated in common areas but strictly forbidden in the saunas. I undressed self-consciously, like the descendant of New England Puritans that I am, then rinsed off in the communal shower.

A sign by the door warned, ominously:

COMPETITIONS TO SEE WHO CAN TOLERATE THE MOST HEAT ARE NOT ALLOWED IN THE SOCIETY'S SAUNAS.

Clutching the little black towel to my privates, I followed Risto into the first sauna, mounting the steep stairs to take our places on the bench (placing the black towel down first, as etiquette requires). An enormous square wood-burning sauna stove occupied almost half the room, crowned by a pile of dark stones that threw off so much heat that it almost had weight to it. The Finns have a word for that, too: *lämpömassa*, or "heat mass." It's a real thing.

During the 1960s, NASA had used these very same saunas to help prepare astronauts for the heat of reentry into the Earth's atmosphere. I was about to find out why. The temperature gauge on the wall said it was well past 100 degrees Celsius (212 degrees F) in there. And it was about to get worse. Risto said some words in Finnish, and I recognized only one: *löyly*. Steam. The other occupants assented. He poured what sounded like a gallon of water into the red-hot rocks, and the superheated steam erupted into the room, tunneling through my nasal passages and down into my lungs. Sweat instantly coated my skin, and my heart began to race. I took slow breaths, trying hard not to panic. I was still very much a sauna novice, and each new session was an occasion for fear and anxiety. But looking around, I saw that a couple of the other guys were hunched over and gasping, so I didn't feel so bad. The Finns call this "*sisu*," and it is analogous to "grit": the ability to tough out a hot sauna session, or any other temporary discomfort.

Most people think of the sauna as a steamy place, but that is not exactly right. In a true Finnish sauna, the heat is usually quite dry, with humidity only around 10 to 15 percent. But when users create *löyly* by pouring water onto the heated rocks, that changes everything. The humidity rockets up to almost 50 percent, making it feel hotter by an order of magnitude as the moist heat hits your skin. In an instant your body becomes completely soaked—every square inch. But it's not all sweat: Studies using special isotope-labeled water have found that the water on sauna users' bodies is roughly 40 percent sweat and 60 percent condensation, as the heated vapor hits their skin, which is likely the coolest surface in the room.[1] Just as sweating removes heat energy via evaporation, *löyly* reverses the process, transferring heat energy *back* into your body and, ultimately, making you feel really, really hot.

On the outside, that is; studies have found that even in a very hot sauna, your internal temperature remains more or less stable, rising by perhaps 2 degrees F (about 1 degree C) over the course of several sauna "rounds," as each individual stay in the hot room is called.[2] But your skin will feel much hotter at times. In this way, the sauna tricks you into thinking you are hotter than you actually are. I did not have Probey to confirm this, but it proved to be a useful mental crutch when the *löyly* was raging and I wanted nothing more than to get the heck out of there. *You're not that hot; you're not that hot . . . really.*

It is considered poor form (and a sign of weakness) to exit the sauna immediately after someone pours water on the rocks, so we sat there suffering until the *löyly* dissipated. Finally, mercifully, Risto stood and headed for the door, muttering, "That's too hot even for me."

Relieved, I followed him to the outdoor terrace, where more than a dozen naked men sat cheek to cheek on benches, steam rising from their shoulders. It was a cold but brilliantly sunny day; the fact that we had all just been in saunas as hot as boiling water made it feel, if not quite warm, at least pleasantly cool.

This being a weekday afternoon, most of the other sauna-goers looked to be past sixty, in some cases well past. Even at seventy-eight years old, Risto was far from the oldest guy there. I felt like I had stumbled into the classic 1980s movie *Cocoon*, in which Wilford Brimley and his fellow elders discover that the swimming pool in their Florida retirement home is a secret Fountain of Youth. The main difference was that instead of dogpaddling around a nice warm indoor pool, these intrepid old Finns were strutting down the metal pier to swim butt-naked in the Baltic Sea, where a bubbler kept the ice at bay.

I worked up my courage and strode confidently down the dock, in full view of the skaters and dog walkers on the frozen bay (in Finland, people hike, ski, and even ride bikes on the winter ice). My swagger, and a few other things, shriveled right up the second my foot touched the frigid water; my leg bones ached like I had dropped a bowling ball on my feet. But there was no turning back. Gasping and moaning, I lowered myself in nipple-deep and stayed there for a hasty count of fifteen. As I leaped back out of the water, an eighty-year-old man came trundling down the ramp and greeted me in Finnish.

"It's nice and warm!" I replied.

"Oh yeah?" he replied with a laugh, in English. "Warm no good!"

He slipped into the water and swam back and forth, as if it were a hotel pool at Disney World.

The Sauna Society is members-only, but exceptions are made for lucky guests and journalists. Joining requires a recommendation from two active members, a modest initiation fee, and the patience to endure a two-year waiting list. For you to get in, a current member must quit, or die—which, possibly thanks to their frequent sauna use, they are generally failing to do.

One wouldn't automatically think that sitting in a room as hot as boiling steam for as long as you can possibly stand it could be "healthy," exactly; other words that come to mind include "brutal" and "punishing." Maybe also "fatal." But a striking series of studies, published over the past decade, suggests that it might be the case. Analysis of data from a large, long-term study of middle-aged Finnish men found that those who used the sauna the most frequently—four to seven times per week—reduced their risk of heart attack, stroke, and overall mortality by 50 percent or more.

The data came from something called the "Kuopio Ischemic Heart Disease Risk Factor Study" (KIHD), which was launched in the mid-1980s as part of a nationwide effort to understand why Finnish males were so *un*healthy.[3] For decades after World War II, Finland had typically ranked among the least healthy or happy countries in Europe, with sky-high rates of alcohol abuse, smoking, suicide, and heart disease. In 1971, Finnish men had among the lowest life expectancies in Europe, largely due to their propensity to drop dead from sudden cardiac arrest.

Finnish life expectancy has since rebounded, to where it now well exceeds that of the US. But back then, things were so bad that a British Army doctor named Colonel H. Foster wrote a pointed letter to the *British Medical Journal* in 1976 speculating that the Finnish penchant for sauna bathing might be the possible *cause* of all these heart attacks.[4]

Finnish doctors and scientists are still offended by that, but on the positive side, it prompted them to begin to look at their native sauna tradition—and their national heart attack problem—in a more rational, scientific way. Beginning in the mid-1980s, researchers enrolled about 2,300 men between the ages of forty-two and sixty who were living in and around the small central Finnish city of Kuopio, about four hours northeast of Helsinki. They then tracked the men for the rest of their lives. (Women were added to the study in 2001.)

The researchers measured the subjects' baseline blood pressure, cholesterol levels, weight, and other parameters and asked detailed questions about their lifestyles: How much did they sleep? What did they eat? How much money did they earn? Did they smoke and/or drink alcohol? And on and on, for about ten pages. Then the researchers tracked the subjects' health "outcomes"—or, in other words, they waited for people in the cohort to get sick or die. The idea was that over time, the study would eventually reveal connections between various lifestyle factors and cardiovascular disease and deaths.

Which it did, but in a totally unexpected way.

In the early 2000s, a Finnish cardiologist and heart surgeon named Jari Laukkanen began working with the KIHD study subjects and their data. At some point, he noticed that the lengthy study questionnaire had also asked participants about their sauna habits: How frequently did they go to sauna? And for how long? This was Finland, after all. It was like asking Americans how often they shower.

Laukkanen's interest was piqued—as a sauna enthusiast himself, as well as a competitive amateur cross-country skier, he suspected that sauna use might improve one's health. But he had no easy way of proving that, short of running a large, expensive clinical trial. As they broke down the KIHD data, he and his statistician colleagues, including his then-wife, Tanjanina, discovered a strong inverse relationship between sauna use and heart attacks—so strong that it practically jumped out of the data at them. The more that the study subjects said they used sauna, it seemed, the longer and healthier their lives had been. "The risk reduction was really, really obvious and clear," Laukkanen has said.[5]

According to their first paper, published in *JAMA Internal Medicine* in 2015, men who visited a sauna two or three times per week were 22 percent less likely to die from sudden heart attacks than those who said they went just once a week, which was a decent-sized effect. But the most frequent sauna users, men who did it four to seven times per week, were more than *60 percent* less likely to die from sudden heart attacks than the once-a-weekers. These frequent sauna users also reduced their risk of overall cardiovascular-related death by half, and their all-cause mortality risk (i.e., their risk of dying for any reason in a given year) by about 40 percent over the course of the twenty-year study.[6]

This striking result made headlines around the world—in part because it suggested that using sauna was almost as beneficial to subjects' health as regular exercise.

Older studies, also mostly from Finland, had suggested that sauna use might have some short-term health benefits, but this study was noteworthy because it spanned more than two decades; it's extremely rare to find such a powerful long-term study of *any* daily behavior. And the enormous size of the effect—cutting all-cause mortality by 40 percent over twenty years—got people's attention. There is no medication that even comes close to that.

Laukkanen and his colleagues have since published more than forty follow-up papers building on these original results. Digging deeper into the data, and cross-referencing it with health and death records (which have been meticulously kept in Scandinavia for centuries), they found that sauna use appears to prevent or improve a wide range of conditions, including stroke—reduced by 60 percent in the most frequent sauna users[7]—as well as hypertension, heart failure, respiratory infections and pneumonia (including Covid), and, perhaps most surprising, Alzheimer's disease. Frequent sauna users had a 65 percent lower incidence of Alzheimer's disease;[8] again, no medication is anywhere close to that effective. Indeed, using sauna seemed to protect people from almost every serious ailment except cancer. They even found that frequent sauna users had suffered 77 percent fewer psychotic episodes over their lifetimes.[9]

In the world of physiology and sports science—particularly in the realm of "environmental physiology," which involves studying the effects of extreme environments on humans—the Finnish sauna studies changed the conversation. "Prior to those studies, everyone was focused on prevention of heatstroke, reduction of heat risk, and how you regulate that," says Anthony Bain, an assistant professor of physiology at the University of Windsor in Ontario. "Since those data came out, I think the focus has shifted to the idea of heat *therapy*."

Instead of looking at heat as something that is always dangerous, even potentially deadly, in other words, the Finnish sauna studies raised a revolutionary question: Could heat be harnessed for healing?

The results suggested it could be—if the studies were valid. After Laukkanen and his colleagues published their first paper in 2015, critics attacked what they saw as flaws in the research, beginning with the important fact that there was no control group to enable a true comparison between sauna users and sauna abstainers. Because the study was conducted in central Finland, where sauna use is nearly universal, the researchers had barely been able to find any men—only twelve out of the 2,315 subjects—who did *not* use sauna regularly. (Nor were there any women in that original sample; women were eventually added to the Kuopio heart-disease study in 2001, and they appeared to derive similar benefits from sauna use.)

So they were comparing people who used sauna a lot to people who only used sauna a little bit—but still much more than the average North American or Western European. I consider myself lucky if I can get to the sauna once a week.

Another issue was that these studies were based on a one-time questionnaire that the subjects had filled out at the beginning of the study; there was minimal follow-up, so the researchers actually had no idea whether these people had continued to use sauna throughout their lives—and if so, how much they used it.

Ultimately, there was no way to say with any certainty that sauna use was the direct *cause* of these folks' apparent better health. It was merely an association, albeit a strong one. Perhaps some of the subjects visited the sauna more frequently because they had more free time than those who could manage it only once per week. "If you're going to the sauna four to seven times a week, that tells me you are less stressed, or more affluent," says Earric Lee, who did his PhD research in Laukkanen's lab and is now at the University of Montreal.

Or perhaps, given that a Finnish sauna round induces a great deal of temporary physical stress, the people who used sauna more often did so because they were healthier to begin with. This is called "healthy user bias," and it complicates all kinds of research in human subjects, notably studies of things like exercise. People who exercise a lot tend to be healthier than those who are sedentary, for example—but at the same time, healthier people also tend to be more willing and able to exercise in the first place. Which is the chicken, and which is the egg?

Similarly, it takes a certain degree of hardiness to withstand a sauna session every day, or even every other day. "You have all these old Finlanders who say they're old because they take sauna," says Mark Timmerman, an American physician who has collaborated with Laukkanen on sauna research. "But maybe they're old because they *survived* all those saunas."

Laukkanen and his colleagues argued that they had already controlled for variables including socioeconomic status, alcohol use, and smoking, and yet the effect remained robust. A later analysis did find that study subjects who had a higher level of fitness *and* sauna-bathed frequently reduced their risk of dying the most (by more than 50 percent); but the less-fit men who used saunas frequently still lowered their mortality risk by a solid 28 percent. So sauna use appeared to confer at least some benefit on its own.[10]

A closer look at the studies reveals that many of these "old Finlanders" were not all that healthy to begin with. They were middle-aged, mostly rural men in what was then one of the unhealthiest countries in Europe. Many of them made abundant use of alcohol, while enjoying minimal access to fresh vegetables for most of the year (prior to the advent of the European Union). More than half of the study subjects reported a family history of heart disease; their average LDL ("bad") cholesterol was on the high side at 156; and nearly one-third of them smoked. Also, rates of stroke in and around Kuopio were sky-high in the 1980s: a reported 322 strokes per 100,000 men ages 25–74.[11] (In the United States, from 2015–19, the stroke rate averaged 58 per 100,000 men.)[12]

By 2011, when the researchers checked up on them again, more than half of the 2,315 men had died from cardiovascular-related causes, including stroke.

So perhaps using sauna wasn't necessarily turning these guys into Supermen but simply preventing them from dying quite so early. But how?

The sauna studies are a perfect example of why epidemiological research—looking at health-related factors across broad populations—can be so tricky in terms of understanding cause and effect. Such studies can reveal tantalizing correlations but cannot "prove" causal relationships.* This is important: Many, if not most, of the studies that you read about on the internet, particularly those involving diet and nutrition, are epidemiological or observational studies, meaning they look at habits or behaviors across large groups of people and try to find correlations—such as that eating exactly twelve hazelnuts per day reduces one's risk of dying (true story). There is no way to "prove" this claim, and it would be equally difficult to run a twenty-year experiment to determine whether using a sauna four to seven times weekly *really*

---

* I'll go a step further and assert that science almost never "proves" anything—our understanding is always expanding and often changing; witness how often long-held dogmas, things we once thought were true beyond doubt, have been overturned. The best we can hope for from any study is that it brings us a little bit closer to the truth. We *could* design a study to determine beyond much doubt whether sauna use prevents heart attacks and heart disease, but that study would require more than a lifetime to run.

lowers all-cause mortality by half. This is why all such findings must be taken with a grain of salt.*

In the 1960s, a British scientist named Austin Bradford Hill—a brilliant epidemiologist who invented the modern randomized clinical trial and was the first scientist to link cigarette smoking with cancer—came up with nine commonsense criteria to help determine whether epidemiological findings are causal or merely coincidental. These are tremendously useful for evaluating almost any scientific research that is reported on in the media; think of them as your own personal nine-point scientific bullshit filter.

I won't bore you with the full Bradford Hill list (see his original paper in the notes),[13] but the Finnish sauna studies satisfied quite a few of the criteria:

1. *How strong is the observed effect?* Laukkanen and colleagues found that frequent sauna users had roughly *half* the risk of heart attack, Alzheimer's disease, stroke, and all-cause mortality as infrequent sauna users. That is a powerful effect, and it means the findings are probably worth paying attention to. Studies with lesser effects, like a 10 percent reduction or increase in cancer risk for some other behavior or food choice, are more open to question.
2. *Does the dose affect the response?* In a causal relationship, the greater the dose of something, the bigger the effect. This held true: The more often study subjects said they used sauna, and the longer they stayed in the hot room, the lower their risk. The people who said they went to the sauna four to seven times per week did better than those who did so two to three times per week, who in turn fared better than the once-a-week people. Not only that, but the men who stayed in the sauna for the longest—more than nineteen minutes per session—also had a much lower incidence of heart disease and heart attacks than those who said they stayed for only eleven minutes or less. This suggests that sauna use may at least have contributed to their better health.

---

* Which, according to epidemiology, may be bad for you.

3. *Are the findings consistent across multiple studies? Have the results been reproduced by other researchers?* Yes and no. Laukkanen and colleagues found that sauna use consistently reduced the risk of death from all sorts of cardiovascular-related causes, as well as cutting the incidence of Alzheimer's disease by as much as 66 percent; a similar effect was also seen in women. So the findings are consistent. But all the highly cited studies came from Laukkanen's research group, which raises questions about their objectivity. Are those researchers perhaps biased in favor of sauna use? Did that affect their findings?

Unfortunately, it seemed unlikely that there would ever be another multidecade study of sauna use and health to compare against Laukkanen's data. But it turned out that there *was* another such study in the works: In 2020, an unrelated Finnish research group examined a completely different and much larger cohort of subjects, comprising nearly fourteen thousand male and female patients who had visited Finnish public mobile health clinics during the 1970s.

The researchers looked at the patients' intake questionnaires, which also happened to ask about sauna use (a common question in Finnish medical histories, apparently), and compared their answers with their long-term health outcomes. They found that patients who had reported using sauna more than twice weekly (or nine to twelve times per month) had about half the risk of developing dementia over the next twenty years, as compared to those who went once weekly or less. This at least appeared to confirm one key finding of Laukkanen's group: that sauna users have a lower risk of dementia. Even better, the effect was observed in both men and women.*

This was promising, but it also added some potential caveats to the story. People who used the sauna at very high temperatures, greater than 212 degrees F (100 degrees C), actually had an *increased* risk of

---

* The *lifetime* risk of dementia among 9–12x/month sauna users was only about 20 percent lower than controls—meaning that some people did develop dementia eventually, but they did so later than the non–sauna users. Even so, a 20 percent reduction in lifetime Alzheimer's disease risk, and a delay in disease onset for another 30 percent, is a solid outcome.

Alzheimer's disease—double that of those who saunaed at 176 degrees F (80 degrees C) or less. More puzzlingly, the most frequent sauna users in this study, the thirteen-to-thirty-times-per-month crowd, saw *no* reduction in their dementia risk, as compared to those who used sauna nine to twelve times monthly. So there was no dose-response relationship. Indeed, this second study seemed to hint that while sauna use may indeed benefit cognitive health in some way, there may also be such a thing as *too much* sauna. (Everything in moderation, as usual.) But it did appear to confirm the other sauna studies, at least directionally.[14]

So we can safely answer that yes, the Finnish sauna results appear to be confirmed by at least some other independent data. But the most important of the Bradford Hill criteria—three of the nine—have to do with *how* a proposed causal relationship might work. What is the likely mechanism? Is it biologically plausible? And can it be replicated or confirmed via direct experiment? These conditions are the most difficult to satisfy, but with sauna use, they also proved to be the most revealing.

What happens when you sit down in a hot sauna—say, at a temperature of 176 degrees F (80 degrees C)?

The Blister Boys would scoff, but for sauna newbies, that level of heat would be terrifying. And, obviously, it is much hotter than the typical gym sauna.

Your skin feels hot almost immediately, especially if someone decides to greet you with a blast of *löyly*. That in turn sets off alarm bells in your brain, as it realizes that you are possibly in danger. Your thermoregulatory system—like the fire department—kicks into action, mobilizing the emergency cooling measures we talked about in the last chapter. Soon you begin to sweat, a bloom of droplets appearing on your forehead, chest, and upper back, where eccrine sweat glands are highly concentrated.

But even before that happens, your body begins adapting to the intense heat. Your heart rate accelerates, sending more blood to your skin and your extremities, your hands and feet; skin blood flow can increase eightfold over normal. Studies by Laukkanen and others have found that after about fifteen minutes in a hot sauna, one's heart rate might easily reach 120 beats per minute, or even as high as 150 bpm, equivalent to a moderate-intensity workout.

Which is why many of the benefits of sauna use are said to resemble the benefits of exercise. Studies by Laukkanen's group have found that repeated sauna sessions, over a period of several weeks, can improve key measures of cardiorespiratory fitness, even without exercising (although the effect is stronger when the sauna is used *after* exercising, if you can handle that).[15]

Even a handful of sauna sessions can improve the flexibility of your arteries, as they expand to accommodate the increased blood flow; it also enhances the functioning of the endothelium, the thin, fragile lining of our blood vessels that is crucial to our cardiovascular health. This may explain why sauna use appears to lower blood pressure—typically knocking about seven to eight points off study subjects' systolic and diastolic blood pressure readings, in patients with early-stage or prehypertension. Even better, the effect lasts anywhere from several hours to a full day. And if you keep coming back, that effect will compound: The long-term Finnish sauna data showed that frequent sauna users were about half as likely to develop hypertension over the two decades of the study.

The improvement in vascular (blood vessel) health that goes along with lowered blood pressure may, in turn, help explain why long-term sauna use also appears to reduce the risk of dementia, which is thought to be caused in part by stiffening blood vessels in the brain.[16] Lastly, Laukkanen found that sauna users typically had lower levels of inflammatory markers, such as C-reactive protein (CRP), which is thought to be a marker for heart disease risk.

If this was *all* that sauna did for you, that would be reason enough to try to sweat it out. Certainly, it got my attention. At my last physical, my blood pressure had clocked in at 145 over 95—well into the red zone, which was a shock because I've always thought of myself as "healthy." On the other hand, this is exactly what people always say about athletic-seeming men (and women, but usually men) who drop dead from sudden heart attacks: "But he was so healthy!"

I went out and bought a home blood pressure monitor, and it too returned consistently alarming results. My doctor muttered about putting me on medication. I didn't want that. Could sauna use help me? And could it help others?

I was convinced that it might—not only on a physiological level but on hormonal, cellular, and even psychological levels. It adds up to an impressive scientific résumé for an ancient practice that, when you get down to it, can be quite uncomfortable, even painful. But the pain, I learned, may be part of the point.

Like exercise, sauna use induces temporary stress on the body and mind. Our response to that stress is what makes us stronger—in cardiovascular terms, as we have just seen, but it goes deeper than that. Finnish scientists have known this for decades. Sauna use strengthens not only the heart but also the immune system, the mind, and even the body. One of the most striking Finnish findings came in a 1986 study that reported that men who used the sauna heavily increased their levels of human growth hormone by as much as sixteen-fold—after three straight days of twice-daily thirty-minute sauna sessions (which is an awful lot of sauna time, even if you are Finnish).[17]

Part of that stress is psychological. The sensation of wanting to get out of there as quickly as possible is modulated by a brain chemical called dynorphin, which accentuates our feelings of distress and dysphoria; this is part of what kicks off our sweating response. At the same time, however, dynorphin helps sensitize the brain's receptors to another chemical that is produced when we are exposed to heat (and sunshine), known as beta-endorphin. Beta-endorphin is sometimes described as the body's own opioid drug, relieving pain and producing feelings of well-being and even euphoria, similar to the "runner's high." Sauna use can produce this sensation without needing to run.[18]

Another way to put this is that sauna use makes us feel bad temporarily, but better in the long run. This cascade of dynorphin and beta-endorphin helps explain why it feels so good to *leave* the sauna.

But while we are in there, the sauna is making us stronger and more resilient, even at a deep cellular level. Studies in organisms ranging from tiny nematode worms on up to humans have found that heat stress activates something called "heat shock proteins," which are molecules in the cell that essentially help to repair and maintain other proteins in our cells.

These are not proteins like the protein in your chicken breast or tofu at lunch. In the cell, "proteins" are like tiny molecular machines, each with a specific job to perform. A cellular protein's ability to do its job depends on its structure, or what scientists call "folding." If it loses its structure or becomes damaged (or "misfolded")—by oxidative stress, ultraviolet radiation, or, of course, heat—that protein no longer functions. (When you crack an egg into a hot pan and the clear fluid turns white, that is the proteins in the egg white becoming misfolded. They are no longer functional; they are food.) Heat shock proteins help maintain the proper structure, or folding, of other proteins in the cell, thus preserving their functionality and helping the cell survive.*[19]

"They're like 'mommy proteins,'" says Simmie Foster, a clinical and research psychiatrist at Massachusetts General Hospital and instructor at Harvard Medical School who has investigated the effects of heat stress on the brain. Those "mommy proteins" nurture and care for other proteins that are "sick" or damaged—even DNA, which, when damaged, can cause cellular dysfunction and even cancer. They also go after bully proteins, breaking up aggregates or clumps of broken proteins that accumulate in our cells, especially around our neurons, where they are thought to be linked to Alzheimer's disease and other forms of dementia.

Emerging evidence suggests that heat therapy—and heat shock protein activation—may have highly specific effects on other key organs, notably the liver. Studies have found that hot water immersion can improve metabolic health in women with polycystic ovarian syndrome (PCOS), for example, via heat shock protein activation.[20] Other research has found that heat therapy improves glucose control, which is important for preventing diabetes; some studies suggest that activating heat

---

* The story of how heat shock proteins were discovered is interesting: In 1962, an Italian scientist named Ferruccio Ritossa left some fruit fly cells in an incubator for too long and they came out looking weird, so he asked some questions about what happened. Heat shock proteins, it proved, were the reason for the structural changes. Heat shock proteins respond to all types of stress, not just heat, but "heat shock" stuck because of the way in which they were discovered.

shock proteins in the liver can help reverse what used to be called nonalcoholic fatty liver disease, or NAFLD (now known, unfortunately, as "metabolic dysfunction-associated steatotic liver disease," or MASLD), a condition of excess liver-fat accumulation that is a key step on the road to metabolic dysfunction and, eventually, type 2 diabetes.[21] Because metabolic dysfunction and type 2 diabetes are such powerful risk factors for cardiovascular disease, cancer, and Alzheimer's disease, this effect may in turn help explain at least some of the long-term health and mortality benefits seen in those Finnish sauna studies.

Certainly, mild bouts of heat stress have been found to help small organisms live longer. In experiments on nematode worms, as well as on fruit flies—simple creatures that nonetheless mimic most of the major systems of the human body—brief doses of heat caused the animals to live significantly longer, also likely thanks to increased heat shock protein activation (and thus, better maintenance of cellular health).[22] Could it do something similar for us?

We don't know. What we do know is that a good hot sauna session leaves us feeling similar to the way we feel after running a 5K: sweaty, euphoric, and somehow purified.

People in Finland will argue to the death that Finnish sauna is the *only* truly authentic and worthwhile form of heat bathing. All other variations are considered somewhat suspect—especially infrared sauna, which (trust me) should never be mentioned in Finland, even in a whisper. (My take: Infrared is better than nothing, and is certainly convenient and widely accessible, but it doesn't really get hot enough to do much of the good stuff that we've been talking about—and there is no *löyly*, if you're into that. It just feels like you are being cooked, which is why it makes me sad.)

Finns also scorn German sauna, which seems virtually identical to Finnish sauna but (according to Finns) is *definitely not* the same. The Finns complain that the Germans impose too many rules on the experience—rules specifying how long one must stay in (15 minutes), what clothing you may wear (none), what you can drink (nothing), and even what you are allowed to say (again, pretty much nothing; silence is golden). In Finland, you can do as you please in the sauna, as long as you're not a jerk. Finns even look askance at the Russian *banya*, despite it (also) being virtually identical to Finnish sauna—just a little steamier and slightly less hot, and involving more beating and whapping with twigs and branches. (Also, possibly, more vodka.)

But in laboratory studies, many other types of passive heat exposure have been found to have nearly identical benefits to the Finnish sauna—broadly speaking—and without requiring a trip to Finland. Even infrared sauna: In Japan, a type of infrared sauna protocol known as Waon therapy has been used to help treat serious cardiovascular conditions, notably heart failure and peripheral arterial disease, with some success. In one remarkable case, a sixty-four-year-old diabetic patient facing possible lower-leg amputation due to a foot ulcer was healed by fifteen weeks of daily infrared heat therapy, which reversed his symptoms and saved his leg.[23]

It doesn't even have to be sauna: A large Japanese study published in 2020 found that people who merely took regular hot baths at home *also* had a significantly reduced risk of cardiovascular death, including from sudden heart attacks. That study followed more than thirty thousand middle-aged Japanese city-dwellers over twenty years, arguably a much healthier population than Cold War–era Finns, and still found a strong effect: People who took hot baths every day of the week, or nearly every day, reduced their risk of cardiovascular "events" and even stroke by more than 25 percent, as compared with the zero-to-twice-a-week bathers.[24] Those results almost exactly mirror the benefits seen in the long-term Finnish sauna studies, though to a lesser degree.

Even the good old California hot tub has been found to have health benefits virtually identical to those claimed for Finnish-style sauna. Chris Minson and his colleagues at the University of Oregon had sedentary people sit in a hot tub for sixty to ninety minutes* and found that it also reduced their blood pressure and arterial stiffness significantly, while improving their endothelial function—the same cardiovascular improvements claimed for sauna use. This may also explain why simply

---

* This is a *long* time to sit in 104.5 degrees F water, so don't try this alone, or without conulting your doctor.

taking hot baths regularly also improves cardiovascular health and reduces mortality risk, according to the Japanese study I mentioned earlier.[25] Other studies, some also by Minson, have found that a good long hot-tub soak (at 104 degrees F) also enhances the effectiveness of endurance training. (Again: Just like sauna.)

The scientific verdict is in: Good things happen when we sweat.

The problem is that we have almost forgotten how.

4

TOO COOL

> Many years later, as he faced the firing squad, Colonel Aureliano Buendía was to remember that distant afternoon when his father took him to discover ice.
> —GABRIEL GARCÍA MÁRQUEZ, <u>One Hundred Years of Solitude</u>

> I've finally accepted that air-conditioning is a privilege and not a right.
> —TED LASSO

HAIRLESS, sweaty, clever, and hungry, our ancestors sallied forth in a world that was, finally, their oyster. Our prehuman forebears acquired the ability to sweat at least eight hundred thousand years ago, long before *Homo sapiens* appeared on the scene. They were essentially tropical primates, ideally suited for running around naked in warm weather. And each successive generation since then, of hominids and humans alike, has lived at the mercy of the weather, in deadly cold and sauna-like heat and everything in between—right up until the mid-twentieth century.

If the three-hundred-thousand-year history of *Homo sapiens*[*] were

---

[*] For sake of argument, we don't know *exactly* when our species first emerged, but it appears to be roughly three hundred thousand years ago, give or take a few tens of thousands of years.

compressed into a single year, modern electric air-conditioning would have been invented approximately three and a half hours ago. Granted, part of this hypothetical year would have been spent in a withering ice age, but even so, one could make a case for dividing human history into two epochs: before air-conditioning and after. It changed our cities, it changed our lives, and it changed us. In many parts of the world, it is almost impossible to imagine what life was like before nearly every indoor space was air-conditioned. And the AC revolution is only getting started.

Today, for example, every hotel and office building in New York City has air-conditioning, without exception—along with 90 percent of all homes in the city (dropping into the lower 80s in some lower-income neighborhoods).[1] Before World War II, however, only public buildings and fancy hotels had cooling systems of any kind; most private residences did not, even in wealthy areas, so people did without. While the rich decamped to the Adirondacks or Newport to escape the dog days of July and August, the working masses sweated out summer heat waves in their tenement blocks—which proved almost ideally constructed for capturing and retaining heat.*

The American playwright Arthur Miller, who was born in New York City in 1915, recalled whole families sleeping on fire escapes at night, which seems dangerous at best. During the day, Coney Island was jammed with people competing for a whiff of sea breeze and maybe a dip in the ocean, if they could elbow aside all the other suffering city dwellers jostling for beach space. If you couldn't get to the beach, you found relief wherever you could, such as in crowded public swimming pools or even on public transit: "Broadway had open trolleys with no side walls, in which you at least caught the breeze, hot though it was," Miller reminisced in *The New Yorker* in 1998. "Desperate people, unable to endure their apartments, would simply pay a nickel and ride around aimlessly for a couple of hours to cool off."[2]

---

* Not to state the obvious, but many older NYC buildings were not designed with central air, and must be cooled via window units that stretch the definition of "air-conditioning." When I lived there in the 2000s, our overpriced fifth-floor walk-up was (barely) cooled by a noisy old window unit that our cheapskate hippie landlord adamantly refused to replace. We solved the problem the old-fashioned way: by opening the bedroom window at night. It would be a little warm at bedtime, but delightful by morning.

Yet Manhattan office workers stoically insisted on wearing suits and ties.

People had been complaining about the heat for centuries—often in places that they were busy colonizing. In colonial Georgia in 1753, Governor Ellis (whom we met in chapter 2) almost had to force himself to sit down to write a simple letter: "For such is the debilitating quality of our violent heats of this season, that an inexpressible languor enervates every faculty, and renders even the thought of exercising them painful," he wrote to his brother back in nice, cool England.

India was even hotter—shockingly, life-threateningly hot. Parts of India rank among the hottest places on earth, and of course the British newcomers insisted on wearing heavy clothes. Overdressed and ill-prepared troops dropped from sunstroke, while doctors scrambled to deal with myriad deadly new tropical diseases and the fevers they caused. There was an urgent need for *any* form of cooling, especially for feverish patients and perishable food.

But ice was hard to come by in these steamy colonial outposts. In the early 1800s, an enterprising Bostonian named Frederic Tudor decided to address that unmet need by shipping ice hacked from frozen New England ponds, packing it in sawdust on fast sailing vessels to India, as well as to the American South and to various Caribbean islands. Ice was prized by physicians during outbreaks of fever, but a good bit of it ended up in wealthy colonials' drinks; indeed, some historians have credited this new global ice trade with birthing the institution of the cocktail. These were huge shipments: The exclusive Byculla Club of Bombay ordered forty tons for the summer of 1840 alone. It made Tudor an extremely wealthy man, and by the mid-1800s he was known as the "Ice King," the undisputed master of the vast and powerful "ice syndicate."

Meanwhile, down in the Florida port town of Apalachicola, on the Gulf of Mexico, a physician named John Gorrie was plotting to put the ice syndicate out of business. Gorrie found himself overwhelmed by patients suffering from tropical diseases like yellow fever, then attributed to "the evils of high temperatures" rather than to the actual cause, mosquitoes. The only treatment that seemed to help his feverish patients feel better was to cool them off in sickrooms chilled by suspended pans of ice. But he was getting pretty darn sick of paying the Ice King's exorbitant bills. So, in 1845, Gorrie quit medicine and began tinkering with ways to manufacture the precious commodity on the spot.

He was, to put it mildly, ahead of his time; the first electric power-generating station (at Niagara Falls, New York) was still half a century away. But it was not completely fanciful. The British physicist William Cullen had accidentally created ice in 1748 by evaporating ether gas inside a vacuum. It was a well-known scientific principle that if you compressed some kind of gas, even air, and then allowed it to expand rapidly, it would instantly get much colder. But creating ice in quantity was a daunting challenge. It took Gorrie a few years, but he ultimately devised an elaborate contraption made up of compressor tanks and pumps and condensers (see the image below) that could freeze large amounts of water. When he finally succeeded, his patients were not the only ones who were thrilled; the French consul in Apalachicola threw a party featuring chilled Champagne—*au revoir* to warm wine!

Patented in 1851, Gorrie's ice machine was a very impressive achievement, considering that the electric light bulb was still decades away. It used the same principles of compression, evaporation, and condensation that are still used to make ice and run air-conditioning systems

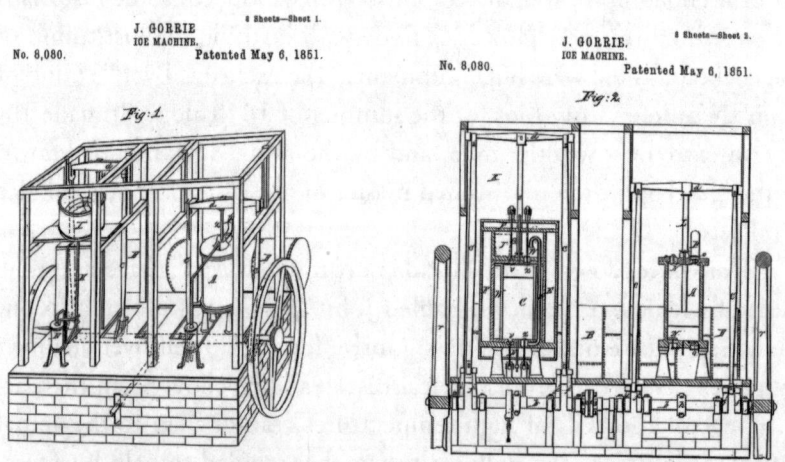

The original ice machine was patented before the light bulb, which gives you a sense of the importance of ice in American life. (US Patent & Trademark Office)

today. Unfortunately, the same year that Gorrie received his patent, his chief financial backer suddenly died, and the business venture collapsed. Broke and devastated, Gorrie himself died a few years later.

History would ultimately prove Gorrie right.* Others building on his idea would eventually put the Ice King out of business, and his invention led directly to the creation of the first air conditioner—by Willis Carrier, in 1902, one year before the Wright brothers' first powered flight. More than a century later, the Carrier Corporation remains one of the world's leading makers of air-conditioning equipment. And Willis Carrier's invention, originally intended only as a way to prevent humidity from ruining paper in a New York printing plant, has proved every bit as world-changing as the airplane.

It took long enough to catch on. In 1940, only 1 percent of US residences had air-conditioning—and not everyone thought they needed it. Even after central air-conditioning was installed in the White House residential quarters in 1933, then-President Franklin D. Roosevelt refused to turn it on, preferring to throw open the huge windows and let breezes circulate through the high-ceilinged rooms—the natural cooling system with which the presidential residence and many buildings of its era had been designed.

Humans had lived for thousands of years in much hotter climates than Washington, DC, often bringing great ingenuity to the problem of staying cool in savagely hot weather. The humble "shotgun shack," popular in poor areas of the American South, was designed to funnel breezes from one end of the home to the other. In certain cities in Persia and North Africa, meanwhile, where summer temperatures regularly reached 120 degrees F, wealthy homeowners built tall, chimney-like towers in order to capture the wind and funnel it down into their living quarters, while also allowing hot air to escape. Many of these wind towers or wind catchers—like natural rooftop air conditioners—still survive, such as in the very-hot Iranian city of Yazd, southeast of

---

* Gorrie is a minor but not forgotten American hero, commemorated by a statue in the US Capitol, as well as a Florida state park, two schools, a museum in Apalachicola, and a bridge—but even that undersells his towering influence on modern life. Can we even imagine a world without ice?

Tehran. Ruins of similar structures have been found in Egypt as well, dating back as early as 3,000 BCE.[3]

In mid-twentieth-century Phoenix, meanwhile, the architect Frank Lloyd Wright—another air-conditioning skeptic—experimented with open-sided, high-ceilinged spaces that would allow cooling breezes to waft over their occupants while still protecting them from the sun. He loathed this new fad for cooling, believing that it was unnatural to go from outdoor heat to indoor chill and back again. "I think it far better to go *with* the natural climate than try to fix a special artificial climate of your own," he declared. "Climate means something to man. It means something in relation to one's life in it. Nature makes the body flexible and so the life of the individual invariably becomes adapted to environment and circumstance."[4]

He was fighting a losing battle.

Growing up in Washington, DC, in the 1980s, I had one or two friends whose families still did not have (or use) air-conditioning in their homes. Always, these were older homes with shady porches, mature trees, and screened-in "sleeping porches" that were quite comfortable. But they were the exception. By 2020, nearly 90 percent of all American homes had air-conditioning of some type, and 66 percent had central air. It is practically mandatory thanks to the way North American homes are now built, with tight windows, low ceilings, little shade, and no more big windows and porches to draw in the cooling breezes.

Yet even now, climate control has only begun to conquer the world. Warming temperatures due to climate change have created demand for air-conditioning in places like Europe, India, and the Middle East that have traditionally been under-air-conditioned in the past. Yet at the same time, ironically, the rise of air-conditioning has also fostered and even exacerbated climate change, to some extent. Cooling our indoor spaces has helped warm up our outdoor spaces, directly and indirectly.

Air conditioner use tripled worldwide between 1990 and 2016, such that cooling our buildings now consumes about 7 percent of the world's electricity production, generating 3 percent of all greenhouse gas emissions. And we are only getting started: The number of air-conditioning units worldwide is expected to triple by 2050, driven by wider availability of air-conditioning—not to mention rising prosperity—in places like Indonesia, Mexico, Brazil, and, especially, India.[5] We're talking about more than a *billion* new air conditioners being installed across the world.

India's AC-related greenhouse-gas emissions alone are expected to jump by a factor of fifty over the next three decades.* These power-generation emissions may not even be the worst contributors to climate change. The refrigerants used in air-conditioning systems are themselves potent greenhouse gases (which makes sense, given that their job is to absorb and transfer large quantities of thermal energy). These compounds, known as hydrofluorocarbons (HFCs), are roughly one thousand to three thousand times more powerful than greenhouse gases like CO_2, and they inevitably leak into the atmosphere when air conditioners are destroyed or serviced.

And yet, who is to say that people in New Delhi or Karachi or London or Athens or Hong Kong don't deserve to have air-conditioning just as much as people in Phoenix or Dallas or Washington, DC? In certain conditions, for certain people, air-conditioning can make the difference between life and death. Or, at a minimum, between comfort and misery.

It's a sticky issue, literally. In France, where hotter summers are driving unprecedented demand for air-conditioning, it has become a political controversy. The organizers of the 2024 Paris Olympics pointedly did *not* build air-conditioning into the athletes' housing, prompting some nations (including the US) to bring their own portable units. French news reports caution that going in and out of air-conditioned spaces can be dangerous to people's health (echoing Frank Lloyd Wright), and that breathing recirculated air is also bad (also true in some cases, as indoor air is often more polluted and more pathogen-filled than outdoor

---

* Of course, like many climate change forecasts, this assumes zero improvements in air-conditioning technology or efficiency over the next several decades, as well as no meaningful increase in power generated by solar, wind, or nuclear sources, or some other low-emission power source that we can't even imagine yet. Nor does it account for a likely slowing of population growth in coming decades. It seems unlikely that humanity will cease innovating or problem-solving around this very personal and increasingly important problem of feeling too hot.

air). On the other side, French right-wing leader Marine Le Pen is campaigning on a promise of air-conditioning for all; meanwhile, the French energy minister Agnès Pannier-Runacher has warned that air-conditioning is "a bad solution" because it will make cities hotter and heat waves even worse.[6]

She is not necessarily wrong. In addition to driving greenhouse-gas emissions, widespread use of air-conditioning also helps exacerbate another climate-related problem, which is the growth of what are called "urban heat islands." As cities have grown larger and denser, they have begun to generate and retain far more heat energy than in the past. And they have become much hotter places than they once were, regardless of climate change.

Atlanta, Georgia, is a perfect example. Once known as "Tree City," it was a verdant Southern town whose streets were lined with tall, shady elms. Those trees were cut down to build skyscrapers, and the surrounding forests and farmlands were subsequently also plowed under to make way for thousands of square miles of new suburbs. As a result, the city center got hotter. A lot hotter.

Back in the late 1990s, a team of researchers from NASA used satellite images and weather-monitoring technology to gauge the effects of these changes to the landscape and found that Atlanta was effectively several degrees hotter than its surroundings; the effect was so dramatic that the city was basically creating its own weather.* The city core was becoming measurably hotter and more humid than its suburbs, with more frequent thunderstorms in the city center than just a few miles away.[7]

More than a quarter century later, Atlanta is even more urbanized, with much more suburban sprawl, than it was back then. In the summer of 2021, students from Spelman College traversed the entire Atlanta metropolitan area on bicycles, carrying temperature sensors, and found that some areas—including the skyscraper-dense city core—were more than 14 degrees F hotter than other, shadier areas of the

---

* It isn't only cities that generate hotter weather. The vast corn and soybean fields of the American Midwest also affect the climate across the eastern half of North America, because the plants produce so much water during photosynthesis. This extra humidity, known as "corn sweat," has been found to worsen heat waves from Florida to Boston. See https://www.washingtonpost.com/weather/2025/07/24/corn-sweat-heat-humidity-east-forecast/.

metropolis.[8] And, as is almost always the case with American cities, poorer neighborhoods tended to be significantly warmer than wealthier areas.

The same pattern holds true in my hometown of Washington, DC, as well as in many American cities built before air-conditioning. The wealthier people fled to higher ground, leaving the poor to endure the oppressive heat and squalor of the urban lowlands. My high school occupied a nineteenth-century estate, on a hill a few miles from the Capitol, that had once served as the "Summer White House" due to its slight elevation and cooling breezes.

Ironically, air-conditioning makes the heat-island effect worse, thanks to the second law of thermodynamics. If you use air conditioners to cool down an office building, all you are really doing is moving heat energy from inside the building to outside the building. That heat must go somewhere, so while the indoors gets cooler, the outdoors gets that little bit hotter.

What's interesting is that this localized heat-island effect is far greater in magnitude than any documented effect of climate change. The climate change debate is focused on an average global temperature increase of 1.5 degrees C, which is of course a big deal, and potentially catastrophic if it continues to 3, 4, or 5 degrees C of warming. But meanwhile, parts of our cities have already warmed up to 5 or 8 degrees hotter (C) than their surroundings simply due to the way they have been built and climate controlled. "Urban heat-island effects are huge," says Patrick Brown, a climate scientist with the Breakthrough Institute, a nonpartisan environmental think tank.

All of this has created a perfect feedback loop: Air-conditioning consumes power that produces emissions, enhancing climate change, which in turn creates wider demand for air-conditioning, particularly among people who live in dense cities. Those dense cities in turn create and capture more heat, requiring still more intensive air conditioner usage, making the urban heat islands hotter still; the heat-island effect likely contributes to the hotter summers that places like Chicago, New York, and even London have experienced.

But change is possible. In New York, green Central Park is often ten or more degrees cooler than highly urbanized, tree-challenged parts of

Brooklyn and Queens. To try to correct this imbalance, then-Mayor Michael Bloomberg launched a massive citywide effort to plant a million street-side trees between 2007 and 2015. The Colombian city of Medellín tried something similar: Once a bleak, hot, violence-torn haven for drug gangs, the city has transformed itself completely over the last two decades, planting nearly three million trees and shrubs that have helped cool summer temperatures by 2 degrees Celsius (approximately 3.6 degrees F).[9]

Another promising initiative took place in Japan, where buildings had been excessively cooled—leading to conflict between male office workers in their suits and ties and female office workers in blouses and skirts. The men wanted it cooler; the women wanted it warmer (sound familiar?). To resolve the conflict, and to save energy and reduce pollution, Tokyo adopted a "Cool Biz" policy in which buildings were cooled only to 82 degrees F—positively sweltering, by some people's standards. But air-conditioning systems were recalibrated to reduce humidity as well, so it felt cooler than that. And dress codes were relaxed, so workers were not required to wear suit jackets and ties all day.[10]

Even as the climate has been changing, as we are constantly reminded—and even as urban heat islands expand and warm up—many people's *experience* of the climate or the weather has been changing in other, often opposite ways. Put simply, while the world has been getting hotter, more of us in the privileged world are spending more of our days indoors, in cool, air-conditioned comfort.

Air-conditioning has changed our very relationship with the weather, bringing an end to untold millennia of humans sweating in the heat. For a growing part of the world's population, sweating has become somewhat optional. We can spend most, if not all, of our lives in a Goldilocks zone between too hot and too cold, where the climate feels *just right*.

Some of this is due to seismic changes in the way our economies and our cities are organized. Around 1900, some 40 percent of Americans lived and worked on farms—meaning they spent much of the day outdoors, in all kinds of weather conditions. The average person would experience a wide range of temperatures in their daily life, without the benefit of heating, cooling, or high-tech clothing. Now, only a tiny percentage of Americans and Europeans are exposed to variable weather in their daily working lives, whether they work in agriculture or in outdoor occupations such as landscaping, product delivery, or construction.

Most of the rest of us spend the majority of our time inside. According to a 2022 survey, nearly 60 percent of Americans spend an hour or less outside per day on average—while nearly 40 percent venture outside for less than thirty minutes a day. Saddest of all is the observation that younger people, aged eighteen to twenty-four, tend to spend the least time outdoors of any generation.[11]

The indoorsification of modern life over the past few decades has helped fuel the rise of the $75 billion antiperspirant and deodorant industry. It is a lot less acceptable to be sweaty if you are spending 95 percent of your time trapped inside with colleagues and loved ones, breathing recirculated indoor air, as opposed to working outside on a farm. Along the way, however, the very act of sweating—even the relatively clean, harmless, and non-stinky eccrine variety of sweating—became stigmatized. Sweat, sweating, and sweatiness are all now decidedly unacceptable outside of the gym, and maybe the sauna. Only air-conditioning could have made this possible.[12]

Air-conditioning is a brilliant invention, a miracle of human ingenuity that, in some places, is all but essential—especially for sleeping in hot and humid climes, but also for survival. Its prevalence tracks inversely with mortality rates in many places, and those of us who have it can consider ourselves fortunate. The ultimate result is that the average modern, rich-world person's daily dose of heat is in fact far *lower* than it would have been a century ago, climate change notwithstanding. If you are reading this book, I would bet that you sweat far less than your grandparents did over the course of a summer's day.

This temporary comfort comes at a physiological cost. Because we are not exposed to hotter weather as much (or colder, for that matter), our ability to handle such extremes has been compromised. That summer heat wave does not merely feel uncomfortable, it almost seems life-threatening, precisely because we are so used to living in an ideal climate. "We're actually working in a tighter range of body temperatures—

like, I'm in seventy degrees all the time," Robert Huggins told me in the KSI lab. "There's inadequate exposure to heat, and so you're not going to be able to adapt when that heat wave comes because we don't expose ourselves to the natural things that we need to help keep us alive."

In practical terms, this meant that I wasn't going to be able to train for the Hotter'N Hell by riding the Peloton in our air-conditioned basement.

"You want to get stronger, you've got to go to the gym," Huggins scolded me. "If you want to adapt to the heat, you've got to go outside."

II
ENDURANCE

5

ADAPTATION

The good Lord made us all out of iron.
Then he turns up the heat to forge some of us into steel.
—MARIE OSMOND

Most years, the Boston Marathon is run on the third Monday in April—Patriots' Day, commemorating the first skirmishes of the American Revolution. In 2021, the race was held in October due to Covid. That wasn't the only odd thing about the marathon that year.

A few miles in, an unknown male runner burst into the lead. This was not unusual. Every year, no-hopers will sprint to the front to showboat for the TV cameras. They never last. But this guy did. It took several more miles for the commentators to figure out that he might be for real.

His name was Clayton Albertson, also known as CJ, from Fresno, California. His only notable finish had been seventh place in the 2020 US Olympic Team Trials marathon, a year and a half earlier. Even that had been a surprise. Now he was leaving the world's top runners in the dust, leading at one point by more than two minutes. He held them off for nearly 21 miles and kept charging even after the favorites chased him down; in the end, he finished tenth, less than two minutes behind the winner, Benson Kipruto of Kenya.

Online, the running world wondered: Who was this guy? And what was he doing at the sharp end of a major marathon?

The answer had to do with Albertson's unorthodox, possibly dangerous, and maybe even crazy training methods.

Just a few years earlier, CJ Albertson had been working as a substitute teacher in Phoenix, freshly graduated from college and unsure of what to do with himself. He had run cross-country and track for Arizona State, but without any earth-shattering results. After graduation, he kept running, not for training, but just for fun. But he got tired of getting up early, so he started running after school—at 3 p.m.—when the temperature in Phoenix could be 105 or 110 degrees F, in the late summer and early fall. Even in that heat, he would still do seven miles every day.

Something clicked. He found he not only liked running in the Arizona heat, but he felt faster and stronger than ever. Once, just for kicks, he cranked out a 30-miler on a blazing afternoon, without drinking any water. He wondered—what on earth is happening to me?

Then he read an online article that answered his questions. Authored by the podcaster, blogger, and PhD scientist Rhonda Patrick, and posted in 2014, even before the publication of the Finnish sauna studies we talked about in the last chapter, the article speculated on all the ways in which "passive heat therapy"—sitting in a sauna or a hot tub—might possibly improve cardiovascular fitness, muscle growth, even brain health. Could heat training, she wondered, be the athlete's new best friend?[1]

Albertson suspected it could. The only problem was that he didn't have a sauna. Or even a gym membership. But he did have a car, and he lived in Phoenix. So he would park his Buick Rendezvous in the sun, letting it heat up to 130 or 140 degrees F and boom: instant sauna. Every day, he would sit in the hot car and sweat, while reading studies about heat training and heat adaptation that he had printed out. "Then I'd be like, I've got to bring a bunch of towels out here because the seats are getting kind of gross," he says. "That was kind of like my first heat training."*

---

* PSA: Don't try this yourself.

Shortly before Covid, he got a treadmill and began running indoors. That changed everything. It wasn't hot enough in his basement, so he rigged up some chicken-coop heat lamps from Home Depot around his treadmill—a homegrown, low-fi version of a heat-training lab. He would heat the room up to 95 or 100 degrees F, boil water on the stove to create humidity, and run and run and run (and sweat and sweat and sweat). His Instagram posts soon went viral, as people wondered what the heck he was doing. "It was just kind of like a challenge," he says. "I would get bored, I think, just running. It seemed hard, so it had to be beneficial."

Indeed: During the pandemic, he ran a 2:09:58 marathon, indoors, on his treadmill, which set a world record. In most major marathons, that would put him with the leaders. He was ready for the next step. As soon as he could, he began entering races again.

After his breakout showing in Boston, Albertson kept notching elite results, with top 10 finishes in Chicago and New York in 2024—within three weeks of each other—and a second place in the California International Marathon in Sacramento. The *Wall Street Journal* dubbed him the "'Mad Scientist' of Marathoning."[2] He didn't win any races, and he didn't qualify for the Olympics, but he achieved something maybe even more impressive: He had morphed from a post-collegiate has-been into a top elite runner, with sponsors and world records—for the indoor marathon, and for 50 kilometers (about 31 miles). Thanks at least in part to the transformative power of heat.

Heat is the bane of the endurance athlete, from the back-of-the-packers all the way to the leaders of the race. To an astonishing degree, more than we realize, heat determines our pacing in endurance sports and even activities of daily life. The hotter it is, the more slowly you will run, walk, bike, or do anything else—even garden.

The reason comes down to supply and demand. Whenever we exercise or do physical work, our heart beats more quickly in order to send more blood and oxygen to our muscles. But if we exercise in a warm environment, our body temperature rises, triggering our thermoregulatory system. That means that we *also* need to send more blood to the skin for cooling and eventually sweating. We can't satisfy both needs. "It sets up a competition," says Chris Minson. "This is why exercising in the heat is so challenging."

This is especially true for runners, who generate a tremendous amount of internal heat yet (unlike cyclists) are not going fast enough to benefit from the convective cooling effect of air flowing over their bodies. The ideal temperature for long-distance running, scientists (and most runners) believe, is somewhere in the 50s F, plus or minus; this is where most marathon records have been set.[3] Any warmer than that, and athletes slow down, sometimes by a lot—and if they don't, bad things can happen, as I saw firsthand on a cool Sunday morning in August of 2024.

I had arranged to join a group of volunteers from KSI in the medical tent at the finish line of the Falmouth Road Race, a classic 7-mile running contest on Cape Cod that draws elite athletes and Olympians, as well as legions of hungover Bostonians, and everyone in between. Huggins, Casa, and about two dozen KSI staffers and students were on hand, as they are every year, because Falmouth is infamous for inducing heatstrokes in runners; back in 1978, the legendary American runner Alberto Salazar had nearly died there of an exertional heatstroke, collapsing on the finish line with a core body temperature of 107 degrees F. He was so far gone that a priest read him last rites.

This seemed like an unlikely day for inducing heatstroke. The air temperature at the start was only around 69 degrees F, and people were walking on the beach in sweatshirts. The humidity is the problem, Huggins explained; that morning it was around 80 percent, making it difficult to impossible for sweat to evaporate. In such conditions, runners can get into heat trouble very quickly, especially in shorter, more intense races like Falmouth or a 10K. This is because there is a lag of about 20 to 40 minutes between when we begin to exercise and the onset of sweating and cooling. We can generate a tremendous amount of heat in that time, and our cooling system cannot offload it quickly enough, at least at first.[4]

Sure enough, as the leading runners arrived at the finish, all hell started to break loose. One of the very first finishers, one of the elite women, collapsed on the line, just like Salazar. She had run herself right into the red zone. Medical volunteers lifted her into a wheelchair and rushed her into the tent, her head lolling and her eyes rolled back into her head. She looked bad.

Huggins and his team quickly took her temperature and confirmed that she was in trouble: her core temperature was 107 degrees F, well into the danger zone. Without wasting a second, they picked her up and lowered her bodily into a Rubbermaid tub filled with ice and water. She was too out of it to protest or even react to the cold. One of the race doctors monitored her vitals, while a nurse called out temperature readings. "106.3 . . . 105.8 . . . 105.3."

Huggins and the doctor relaxed: She was headed in the right direction, albeit gradually. A few gallons of water, a $200 tub, and a couple of 20-pound bags of ice had made the difference between life and serious injury or even death. Meanwhile, about a half-dozen other runners had been brought into the tent with elevated body temperatures, including an intense man in his thirties who had been trying to beat his personal record, and a cocky college bro who had sprinted up the last hill—"Tryin' to be Supahman," as he put it. Supahman, too, ended up lying in a tub of ice water with a temperature probe in his rear.

The elite runner recovered quickly. Within ninety minutes, she was able to leave the tent under her own power, though she still seemed a little unsteady. The only person who seemed disappointed was her manager, who had hovered anxiously the whole time. "This wasn't the outcome we were hoping for," he muttered. But it could have been worse.

Three decades ago, the South African sports scientist and runner Timothy Noakes wondered: Why don't *more* runners suffer heat illnesses in warm-weather marathons? Or even die? Why are deaths and even heatstrokes so relatively rare, in marathon running? Even on hot days?

It was a simple question with profound implications for our understanding of endurance physiology. For decades, sports scientists had assumed that endurance athletes were little different from automobiles: They could only go as fast, and as far, as their engine and fuel supply would permit. In other words, their performance depended entirely on their aerobic capacity, their body's ability to supply oxygen and fuel to their muscles.

Noakes felt that this purely mechanistic model did not explain what happened in the real world. Why could some runners summon the energy to sprint at the end of a long race? Shouldn't they be slowing down? And yet, at the exact point where they should be close to "catastrophic failure," according to the standard, car-engine model of performance, they were able to go even faster. At the same time, other athletes sometimes appeared to underperform, going slower than predicted by their past performances or measures such as VO_2 max.

Why didn't the fittest athletes always win?

Because athletic performance is not only about physical capacity or fitness, Noakes realized. The size of our "engine" matters, but the throttle is controlled by something else: our brain. The brain decides—often unconsciously—when we can go faster, and when we must slow down, lest we go too hard and get ourselves into trouble. Noakes called this mysterious mechanism the "Central Governor," and it remains highly controversial—for one thing, its many critics note, it is impossible to observe directly or disprove via experiment. But it can nonetheless serve as a useful model for thinking about athletic performance.

One of the most important limiting factors to the Central Governor, Noakes realized, is heat. When the brain senses that we are getting too hot, it puts on the brakes—even if we *want* to go faster—to keep us out of danger from heat exhaustion or heatstroke.

One of Noakes's graduate students, Ross Tucker, did an elegant experiment demonstrating how this might work: He found that cyclists in a heated chamber rode more slowly than they did in a cooler chamber. This was not surprising in itself; what *was* surprising was that they slowed down from the very beginning of the hot ride, before their body temperature had risen even a tenth of a degree. Further testing revealed that they activated fewer leg-muscle fibers in the heat than they had in an identical ride under cooler conditions. This suggested that their brains, sensing that it was dangerously warm, had quite literally shut down their "engines."[5]

"The performance impairments happen before you start," Tucker told me. "The brain goes, *hold on, pace yourself.*"

Which happens to athletes all the time, often to their surprise and disappointment. For example, my friend Sylvia Bedford, an elite marathon runner from Utah, entered the 2024 US Olympic Trials marathon hoping for a solid finish. It was her second Trials race, and although she had no ambitions of actually making the US Olympic team, she had won smaller races in the Mountain West and looked forward to running a competitive race. The trials were held in Orlando, Florida, home of Disney World, and even though it was February, race day was fairly warm (for running), with temperatures in the upper 60s Fahrenheit and typical Floridian high humidity.

Sylvia felt fit and well-prepared. She relished this kind of high-level competition and hoped to push herself. But she found herself struggling right from the start. "Normally you can just sort of lock into your pace and go," she told me later. "But I looked down at my watch at mile 4, and my heart rate was redlining already. I felt like I was working so hard to hit my usual pace."

She ended up dropping out at mile 19, feeling dizzy, and sat down on the curb. She was not alone; 20 percent of the field quit before the finish. But it didn't make sense. She couldn't come close to what she had done in training.

Her problem, she eventually realized, was the heat. She had done most if not all her training in cooler fall and winter weather in Northern Utah. Yet a relatively mild day in Florida had completely stymied her. Although she was extremely fit, she was not prepared for the high heat and humidity. And the Central Governor had done its job—saying, *hold on, pace yourself.*

Sylvia was lucky: Athletes who disregard their Central Governor, or try to go beyond its limits, can get in serious trouble. The British cyclist Tom Simpson was vying for victory in the 1967 Tour de France, for example, when he collapsed near the sunbaked summit of Mont Ventoux, one of the steepest, hottest, and most feared climbs of the race. He died from heatstroke, right on the roadside. It later turned out that he had taken amphetamines before the race start, washed down with brandy, which likely short-circuited his Central Governor.[6]

For non-elite athletes, and especially for the sort of average runner who might complete a marathon in three and a half to five hours, the Central Governor is even more strict. Recreational runners, people who typically run 7:30- to 10-minute miles, slow down drastically in the heat—by about four to four and a half seconds per mile, or more, for each additional degree Celsius of warmth, according to one study.[7] Elite runners slow down too, but by only about one second per mile per degree C. Even so, the winners' times for warmer races are often not that much slower than for cooler marathons. Because, for both elites and amateurs, the limits of performance in the heat turn out to be highly flexible.[8]

To put it another way: We can often do more than we think we can, and heat training can help us stretch those limits.

More and more, the road to athletic greatness runs hot—certainly in endurance sports such as cycling, distance running, and triathlon, but also (as we will see) in team sports like American football.

Professional cyclists have been using heat training and heat-adaptation methods for years—deliberately training in hot conditions to prepare for increasingly hot weather during major summer races like the Tour de France, held in July; temperatures in the mid-90s F and higher, once rare, are now common. As cycling has expanded beyond Europe, with important events held in Australia and the Persian Gulf, the need for heat adaptation has only become more obvious. But heat training has turned out to have broader benefits as well.

The heat-training revolution began with a handful of outsider athletes working mostly in secret or even alone, like CJ Albertson.* Now, heat training has become almost mandatory. The 2020 Tokyo Summer Olympic Games marked the turning point. As soon as the venue was announced, it was clear that Tokyo would be one of the hottest, if not the hottest, Olympic Games ever held—hotter than Rio de Janeiro, Athens, or Atlanta. Tokyo is warm and also very humid in the summertime—even more so thanks to climate change in recent decades, coupled with the heat-island effect of being a city of 40 million inhabitants.

But the full extent of the toll that the heat would take was not fully clear until the triathlon "test event," held on the Olympic course in central Tokyo in August 2019, a year before the scheduled Games. It was so hot leading into the event that the women's run leg had to be

---

* The first professional athletes to employ heat-training techniques were probably old-school boxers, training in un-air-conditioned gyms in Philadelphia or New York City, and running in heavy clothes and even trash bags to sweat out water weight, which likely also added some heat-acclimation effects to their fitness.

cut in half, from 10K to 5K. The water temperature on race day was 86 degrees F, or about what one would expect in a senior-center pool, not an elite athletic venue. Many race favorites, male and female, were unprepared for the hot, humid weather, recalls Olav Aleksander Bu, head performance coach for the Norwegian team.

"Some people basically came as late as possible into the test event from Europe, and their strategy was to not stay outdoors at all before the race start," he told me. "They were just delaying everything, absolutely to the latest point. And one of the reasons that I heard was that they wanted to avoid sweating."

Bad plan. As the race unfolded, the favorites crumbled one by one, especially on the run; fully one-quarter of the men's field did not even finish. But Bu had been training his Norwegian athletes for hot conditions since 2018, using protocols based on US and Israeli military research. Because Norway is not very warm, he had his athletes ride and run wearing multiple excess layers of clothing, making themselves as hot and sweaty as possible, while monitoring their core temperature via special, experimental devices. It was miserable, but it worked. Despite coming from a cold climate, the Scandinavians triumphed: Two Norwegian athletes, Casper Stornes and Gustav Iden, finished second and fourth in the test race.

Heat training for performance had been something of a fringe idea until then, at least outside of cycling. After Tokyo, it was no longer optional at the elite level; it was mandatory. The year-long delay of the Games due to Covid not only gave athletes a break from competition but allowed teams to experiment with different training approaches. By the time the Tokyo Games were held in August 2021, many athletes across a wide range of endurance sports—from race-walkers[9] to triathletes, and even dressage horses[10]—had incorporated some kind of heat-adaptation work into their training. And the Norwegians triumphed again: Kristian Blummenfelt won gold in the triathlon, while Iden won bronze.

I was no Olympian, nor even a dressage horse. But I wondered—could heat training get this slow horse running again?

The morning after my first heat-lab session, I woke up hungry and wolfed down a hotel breakfast burrito. My body was still wrecked from all that sweaty pedaling. It felt like a four-star hangover. Filled with dread, I dragged myself back across campus to the KSI heat lab, to do it all over again. Hair of the dog, and all that.

Today, Huggins had promised, we were going to work on heat acclimation. Whether you are an Olympic triathlete or a South African gold miner, or a mediocre recreational athlete like me, heat acclimation looks much the same for everyone: You work hard in the heat for an extended period, until your body temperature rises to a certain level. And then you do it again, and again, and again. Which sounded terrible, but Huggins made clear that it would not be optional.

"Why can't I just work on my fitness?" I had asked him the previous day. After one session in the heat chamber, the idea of heat training, outside, seemed quite unappealing. "That ought to count for something," I whined.

It doesn't, he said. He had watched many "fit" people completely fall apart at the Hotter'N Hell and other hot endurance events. There was no getting around the need for heat acclimation, or as the cool kids in the lab called it, "heat acc." After my disastrous first day in the heat lab, he knew it was probably my only hope.

"I want you to be safe when you go to Wichita Falls," he had said, looking genuinely worried. "It's no joke."

Pouting, I changed into my bike clothes, got reacquainted with Probey (easier the second time), and remounted the dreaded Velotron bike with the DUI handlebars. There was no getting out of heat acclimation.

Huggins had put the heat chamber on blast, even hotter than the previous day, with no fans this time. As I started pedaling, he tapped his phone and grinned as the speakers blared the Foreigner song "Cold as Ice."

This workout promised to be even less fun than yesterday's. The goal, Huggy Bear explained, was to get my core body temperature up to 38.5 degrees C (101.3 degrees F), the level where studies say that physiological heat adaptation begins to occur—including heat shock protein activation[11]—and keep it there for an hour.

I could have wept. *An hour?*

The only way to develop heat tolerance, he explained, is to get hot, and stay hot. Just not *too* hot, and not for too long. (Which is why we required the services of Probey again.) Easier said than done, it turned out. Yesterday, it felt like I was flirting with heat stroke in next to no time. Today, it seemed to take forever. Although the room was quite toasty, and although I was pedaling hard on the bike, my core temperature refused to budge—thanks to good old thermoregulation.

Sweat gushed from my pores, my heart raced, and my tongue lolled out like Bode the dog's on a summer afternoon. At one point, a gaggle of prospective students filed into the heat lab on a campus tour. Their innocent young jaws dropped in horror at the sight of a pasty-looking man their dad's age in a sweat-soaked jersey and too-tight bike shorts, looking like he was about to have a heart attack right in front of their eyes. They filed back out in traumatized silence.

Even after 30 minutes of sweaty misery, my temperature stubbornly remained below 100 degrees F. This was reassuring, in a way; I would probably make it through at least the first half-hour of the Hotter'N Hell Hundred. Less reassuring were the noises coming from my stomach as it dueled with the breakfast burrito.

Finally, after 52 minutes of ever-queasier pedaling, the probe reached the magic temperature of 101.3 degrees F. I was cooked. Now came the fun part of heat training: Doing nothing. Huggins let me hop off the bike and just sit in a chair sweating for 10 or 15 minutes, resting like a medium-rare rib eye fresh off the grill. The playlist served up Foreigner again: *I'm hotblooded, check it and see* . . .

When my temperature started to drop, I got back on the bike and pedaled some more. Then I rested again. And pretty soon, I was done. And that was heat training: You work until you get hot, and then you just kind of hang out. Simple as that. Because just *being hot* is enough to trigger the physiological adaptations that make a person more heat tolerant.

I would need to do this heat-based workout about eight or ten times in the lead-up to the Hotter'N Hell, Huggins explained, later in his office. If I wanted, he added, I could also get into a hot bath or a sauna after riding. That sounded like the worst possible idea in the moment, and my lack of enthusiasm showed.

He sighed and stared into his laptop screen.

As my friend Sylvia learned at Olympic Trials, aerobic fitness and heat adaptation are two different things. Improving fitness brings some partial heat adaptation, almost as a side effect. But true heat adaptation requires specific training. Just as athletes must train their muscles and cardiovascular system, Huggins explained, I would need to also train my innate cooling system.

David Martin had agreed to coach me—and, I suspected, serve as Huggins's spy. A semipro triathlete and ultrarunner himself, David is a cheerful, upbeat guy who probably soon regretted saying yes to grouchy, lazy me. After I came home from Connecticut, he designed a training plan that would help me prepare for the Hotter'N Hell.

I dutifully ignored it. I meant to start training, I really did, but life kept getting in the way. The Fourth of July came and went. The Hotter'N Hell was in late August. Time was ticking.

Most people tend to be more active and fit in summer than the rest of the year. I am the opposite. I dislike exercising in the morning, preferring to drink coffee and walk the dogs, then do a little work as I ease into the day. By the time I'm ready for a break, it is usually too hot to go for a bike ride or a hike or do much of anything else outdoors. So I'll make a big fat sandwich instead.

My lazy summer routine turned out to be perfect for heat training. I could putter around and work for a few hours, *then* go out and train. I was supposed to wait until it got hot. That was the plan, anyhow. But the first couple of heat-training rides were wretched, zombielike affairs, as I turned the pedals and sipped disconsolately from lukewarm water bottles. I felt like I was dying.

But I wasn't; not even close.

I knew this for a fact, thanks to a nifty little device called a CORE that measured my body temperature as I rode. A one-inch plastic square that clipped to a chest strap, the CORE measures "heat flux," the amount of energy flowing *through* the wearer's skin, via a complicated (and proprietary) set of algorithms and equations.[*] Then it considers skin temperature and heart rate, and calculates core body temperature from there.

Complicated as it sounds, the CORE was actually fairly accurate. I had worn one during my sessions in the KSI lab, and it had tracked the rectal probe pretty closely, without requiring the use of Vaseline. If

---

[*] The CORE is based on technology developed in the aerospace and laser industries, intended to measure the heat load inside very complex and expensive machines.

it was good enough to be used by Tour de France cyclists and professional triathletes, including Bu's killer Norwegians, it was good enough for me.

Anyway, according to the CORE, I was nowhere near dying. In fact, I needed to get hotter. According to David Martin and Huggy Bear, I was supposed to ride long and hard enough, in hot enough conditions, to get the CORE reading to at least 101 degrees F—and keep it there, ideally for an hour. Easier said than done, it turned out. There is a popular road climb near my house called Emigration Canyon that rises about 1,300 vertical feet over seven and a half miles, up into the foothills of the Wasatch Mountains, so that's where I went. It seemed like that should get the job done. But even on the most sunbaked summer afternoons, it took even longer to get hot on Emigration than it had in the heat lab, probably because sweat evaporates almost instantly in Utah's 15 percent humidity. Eventually, the CORE would hit 101, and I could ease off and just sort of gently bask in the sun. It wasn't so bad.*

I soon became obsessed with the CORE. It was revelatory—like wearing a heart-rate monitor for the first time—because it opened a new window for understanding physiology and performance. Instead of relying only on heart rate, I could now track my heart rate *and* temperature. I began using it for nearly every ride, watching my body temperature rise and then stabilize. The CORE calculated a "heat strain index," measuring the thermal stress of each workout, and I could use that to chart my progress toward heat acclimation.

While riding in the heat felt uncomfortable at first, the CORE provided a certain level of confidence. However hot I might feel, the device reassured me that I was not at imminent risk of heat illness. Not even close. This, weirdly, was the greatest gift, because over time, it enabled me to recognize the difference between slight discomfort and actual danger.**

After five or six of these "heat acc" rides, I started to notice changes.

---

* Later I discovered that I could fairly easily get the CORE temperature up close to 101 F on a hard 45-minute indoor ride on the Peloton, but I didn't know this at the time

** Minson suggests a different, more intuitive (and cheaper) way to gauge heat stress while training: By feel. If 10 is the hottest you've ever felt, and zero is perfectly comfortable and cool, he suggests that athletes try to maintain a self-perceived heat-stress level of 5 or 6 out of 10. I found that it was easier to gauge my level of heat stress *after* I had used the CORE for a while and had learned to recognize the difference between feeling a little bit hot and feeling dangerously too hot.

I no longer minded the heat so much. My muscles felt looser and more relaxed, and the sun felt good on my skin. Even better, I had the local trails and roads to myself, because almost nobody else was dumb enough to go out in the mummifying heat. As August wore on, I began to feel stronger and more confident in the heat than I ever remembered. I was actually looking forward to the Hotter'N Hell.

The biology of heat adaptation is simple yet elegant. As we exercise or even just sit around in the heat, our physiology begins to change in subtle ways. This is why, as noted earlier, a warmish August day is so much easier to handle than the exact same temperature in May or June. Heat adaptation is a fancy way of saying we are getting used to the heat.

After just four or five one-hour sessions of heat exposure—which could be nothing more than taking a walk outside on a hot day—studies have found that a person begins to become more tolerant to the heat. After ten to fourteen sessions, most people become fully adapted. This method had worked for the South African gold miners back in the 1930s, and workers and athletes have been heat-training in pretty much the same way ever since. It doesn't need to be complicated or time-consuming.

One of the most important changes that takes place as we get "used to" the heat is that our blood plasma volume begins to expand. (Plasma is the liquid component of blood, as distinct from red or white blood cells, cholesterol-carrying particles, hormones such as insulin, and everything else that is transported via the blood.) This means we now have more blood in our circulation, to satisfy the competing demands of cooling (sending blood to the skin) *and* movement (supplying oxygen to the muscles and the brain). It takes a few days of heat training for this process to begin, but in people who are fully heat-adapted, plasma volume can eventually increase by as much as 15 to 20 percent, or about a liter.

This is crucial because one of the biggest changes that comes with heat adaptation is that we sweat more, and sooner.[12] This seems somewhat counterintuitive: shouldn't heat acclimation mean we are somehow resisting the heat and sweating *less*?

Quite the opposite: Heat adaptation turns us into super-sweaters. For starters, we sweat more. Our sweat glands are activated more easily, so they begin to sweat sooner and more copiously. More plasma volume means that we have more liquid available to turn into sweat. At the same time, our sweat becomes less salty.

This is one of the most interesting pieces of heat adaptation: While we think of sweat as salty, it is less salty than our blood or the fluid between our cells. This is because our sweat glands are equipped with an ingenious little mechanism that removes or reabsorbs sodium and potassium ions from sweat before it reaches the skin surface. This conserves precious electrolytes—imagine how rare salt would have been, in hunter-gatherer times. When we are heat-adapted, this reabsorption mechanism kicks into overdrive, sucking still more sodium from our sweat, which enables that sweat to evaporate more easily and cool us more efficiently. One study found that sweat salinity begins to decrease after just two days of heat-acclimation training.[13]

This super-sweating is not only important for cooling the body core, but the skin itself. Having a cool(ish) and more importantly *stable* skin temperature essentially fools our brain into believing that we are OK and that we can continue safely—even if, as is often seen in elite athletes, our core temperature remains relatively and even dangerously high. This is why elite marathoners finish races with core temperatures of 104 degrees F or higher, well into the danger zone for heatstroke, yet without any apparent problems. Because their skin stays relatively cool, their Central Governor permits them to continue.[14]

Some researchers believe that keeping a stable, relatively cool skin temperature may indeed be a primary factor in determining performance—trumping core temperature *or* hydration status (which we will discuss in the next chapter).[15]

Notice how these adaptations parallel some of the changes that we have talked about in long-term sauna users. Both activities involve intensive sweating, an elevated heart rate, a slightly increased core body temperature, and no small degree of cardiovascular stress. This is no accident. And there does seem to be a crossover effect. Studies have confirmed that athletes who use sauna *after* their workouts, like CJ Albertson or US Olympic marathoners Conner Mantz and Clayton Young, and others, improve their endurance.[16]

"That can kind of extend your workout, if you're not feeling like going for a full run that day," Albertson told me.

For a long time, scientists and coaches believed that heat training *only* helped athletes cope better with the heat, and that it was not useful in other ways. Thus they tended to use heat training sparingly, because it is so physically and mentally stressful for the athletes. (True!) But gradually, Minson and other elite coaches such as Bu had noticed that their heat-trained athletes also seemed to perform better in cooler conditions. Heat training appeared to improve performance across the board. Why?

It turned out that long-term heat adaptation causes deeper changes to an athlete's physiology. Over four to five weeks of heat training, athletes not only expand their blood plasma volume, but also increase their hemoglobin mass—the oxygen-carrying protein in red blood cells.[17] This is a big deal: It means that their blood could transport more oxygen to their muscles, so they could do more work for longer. It was similar to the effects of altitude training—or illegal blood-doping, for that matter—only it was much easier and cheaper to do. And while heat training didn't seem to confer quite the same magnitude of benefit as either altitude training or blood doping, it had the advantage of being completely legal.*

In the ultracompetitive world of elite endurance sports, this was already an open secret, known among a handful of coaches but not yet published in the scientific literature. Finally in 2020, a group of Norwegian scientists published a study showing that elite cross-country skiers and cyclists who had heat-trained for five weeks, wearing extra layers of clothing to make themselves hotter, increased their hemoglobin mass by nearly 5 percent.[18] A big deal. The extra heat stress had potentiated their training, making their workouts more stressful yet more effective, unlocking new levels of performance.

In this, it is comparable to altitude training, which also improves performance slightly but definitively.[19] But heat training may improve

---

* Luckily for me, my Utah bike rides counted as both altitude and heat training, reaching elevations as high as 6,000 to 9,000 feet above sea level.

performance for other reasons as well: One interesting theory posits that the performance benefits observed in heat-trained athletes may in fact stem from increased heat shock protein activation.

The magnitude and mechanisms behind heat-training performance benefits remain controversial, with some researchers dismissing them as minuscule and others like Minson going all in. Minson estimates that heat training can help his elite athletes improve their overall performance by about 2 percent, which is significant in high-level competition; for amateurs, however, the potential gains are much greater, and certain athletes can benefit disproportionately. "We get some who do have big improvements, and I think usually those are the ones who 'don't like the heat,' and they were avoiding the heat," he told me. "Those individuals will see a much bigger increase—it's pretty amazing, actually."

That was me! I was one of those athletes who "don't like the heat"—*and* I am always looking for shortcuts to avoid hard work and unnecessary suffering. Yet, as I would learn, the benefits of heat training bring not only physical improvements, but mental ones as well.

"When someone gets adapted, we can see the physiology that changes," Minson told me, "but one of the most amazing things to me is just how much more of a comfort level they have."

As the summer went on, I tackled my heat-training assignments with renewed gusto. David Martin, my coach, began sounding more hopeful in our weekly calls.

I quickly lost my long-standing fear around riding my bike or even trail running on hot days. Now I was one of those weirdos you see out exercising when they probably shouldn't be. What I didn't really understand yet was that heat training was doing far more than making me a better athlete. I was improving in ways that had little to do with athletic capacity at all—ways in which heat training and heat adaptation make us stronger and more resilient overall, mentally as well as physically.

With every ride, I felt more comfortable in the heat, and less afraid of it—a change that turned out to have a basis in biology. Remember those heat shock proteins, which we talked about in chapter 3? Sustained bouts of heat exposure, such as heat training, activates our heat shock proteins—to a greater extent than sauna—and keeps them activated.

And while they may do good things for our health, their main job is to protect us from the dangerous effects of heat stress itself. The more heat shock proteins that are activated, the more stress-resistant we become, and over time, this cellular resilience accumulates. Experiments done in a US military lab found that when mice were exposed to severe bouts of heat stress, repeatedly, their cells themselves became more resistant to damage from the heat—in part thanks to increased heat shock protein activation. This in turn made the animals better able to withstand the adverse consequences of heat stress, up to and including heatstroke.[20]

In other words, every hot training ride that I did—keeping a careful watch on the CORE, to make sure my core temperature did not get too hot—was making me less susceptible to heat illness.

Prompted by concerns about climate change, scientists have studied heat stress not only in humans and mice, but in a diverse array of creatures including cattle, thoroughbred horses,[21] dogs, yaks, and Chinook salmon,[22] to name a few. In most of these creatures, heat stress and heat adaptation not only lead to better tolerance of weather extremes, but also profound cellular resilience. One interesting finding has been that heat shock proteins do more than protect us from heat. Studies have found that heat acclimation can fortify mice (and presumably other animals, possibly including humans) against other kinds of stress and damage, up to and including traumatic brain injury.[23]

Another important point is that heat adaptation is not only helpful for athletes; in fact, it may be even more important for people who are less fit (and thus more vulnerable to heat stress). In one recent study, Chris Minson and his colleagues had sedentary volunteers sit in hot tubs until their core body temperature reached the magical level of 38.5 degrees C (101.3 degrees F)—the level at which most heat adaptations take place. They then maintained that temperature for 60 minutes. After eight weeks of doing this four or five times per week, all the subjects had lowered their blood pressure, improved their arterial flexibility, and enhanced the function of their endothelium (the lining of their blood vessels, again) in a way similar to the effects of frequent exercise (or, for that matter, frequent sauna use).[24]

"The magnitude of improvements in vascular function and blood pressure observed in the present study was similar to what is typically observed in young, healthy, sedentary subjects with exercise training," the researchers wrote, "and in some cases, even greater."[25]

In yet another study, a mild exercise program coupled with some heat acclimation helped heart patients recover more quickly from bypass surgery than those who had not heat-trained prior to going under the knife. All of this strongly suggests that heat training is not just for athletes; it has therapeutic potential that is only beginning to be explored.[26]

By early August, I was becoming convinced that heat training might actually help me, a decidedly average athlete, get through the Hotter'N Hell Hundred.

6

DRINKING PROBLEM

> Water is the driving force of all nature.
> —LEONARDO DA VINCI

AFTER my second very hot session in the KSI heat chamber, I still wasn't done. Things were about to get weird; it was time for the Sweat Test.

Langan and Martin, the triathlete lab assistants, bundled me out into the corridor and had me squat down in a black plastic tub, the cord of Probey dangling awkwardly from my bike shorts. Huggins tossed in the sweaty towels I had used, along with my soggy jersey. It felt like I was about to get whacked, like an informant in *Breaking Bad*.

"Time to wash you down," Martin announced. Hoisting a plastic gallon jug, he began slowly pouring distilled water over my head and shoulders.

"Scrub your hair," he commanded. "Scrub, scrub, scrub!"

I scrubbed.

"Neck! Shoulders! Armpits!"

Nobody batted an eye at the sight of a half-naked man crouching in a kiddie pool in the hallway, getting washed down by three dudes wearing blue latex gloves. "It's like the most awkward bath you've ever had," Huggins joked, as 90 minutes' worth of sweat was rinsed away in public, along with the last shreds of my dignity. I didn't mind; the cool water felt delightful, as it flowed over my skin and pooled in a cloudy-looking broth in the bottom of the tub. "This is the sweat soup," Huggins said.

In most labs, the PhD scientists who get the grants and publication credits and awards never get anywhere near the body fluids of any human or animal study subjects. Their underlings do the wet work. But Huggins dove fearlessly into the sweat soup, because he is that kind of guy and because it is almost impossible to gross out a hockey player.

By analyzing the sweat soup, Huggins and his staff could determine exactly how much liquid I typically lost by sweating during exercise in the heat, and how much salt and other electrolytes such as potassium and chloride that sweat contained. In other words, how salty was my sweat? This information would be used to develop a personal hydration plan, telling me what to drink during the six or seven (or eight or nine) hours it would take to finish the Hotter'N Hell Hundred, in order to replace the fluid and electrolytes that I would lose while riding.

Hydration is the flip side of sweating; we need to replace the fluids we lose while cooling ourselves in the heat. Obviously. As straightforward as this sounds, however, the whole topic of hydration turns out to be fiercely contentious. Scientists cannot agree on basic questions, including how much liquid athletes (and nonathletes) really need to drink while exercising, what liquids they should be drinking, when they should be drinking them—and why.[1]

Adding to the confusion is the fact that, on many key issues pertaining to hydration, the conventional wisdom has been completely reversed, within living memory. Not that long ago, athletes were warned not to drink anything at all while exercising. Now at a soccer game or a running event, nearly everyone—spectators and participants alike—seems joined at the lip to an enormous steel water bottle for fear of becoming dehydrated. The internet and social media are buzzing with hype about hydration and health—check the TikTok hashtag #GallonChallenge—to the point where some scientists are beginning to worry about the dangers of *over*hydration, people drinking too much water.²

Sound crazy? It's not. Overdrinking nearly killed Brooke Shields a couple of years ago. More on that in a minute.

Meanwhile, hydration has grown into an unbelievably lucrative industry—witness the vast drinks section in any convenience store—making it tricky to separate truth from marketing.

In the hot, dry American West, there's an old saying: *Whiskey's for drinkin', but water's for fightin' over.*

In the world of sports science, hydration is for fightin' over. And the stakes are higher than you might think.

Water, ultimately, is what enables us to thrive in the heat—the constant flux of water into, and out of, our bodies. We need to drink water in order to survive. But rampant misinformation and even manipulation of the science has muddied the truth so much that it can potentially threaten our health—and even kill us, as in the tragic case of a Boston Marathon runner named Cynthia Lucero.

Running was Lucero's passion, both personally and professionally. She jogged for fun and fitness, but she was also a clinical psychologist who had written her doctoral thesis on the benefits of marathon training for the family members and friends of cancer patients. A few days after her thesis was published, in April 2002, she got to run the Boston Marathon. It was, she told friends, a "week of triumph."³

It was only her second marathon, so she took it slow and steady, chugging along at a pace of about 15 minutes per mile. The weather was cool that day, with a high of just 52 degrees F (11 degrees C), but it appears that Lucero stopped at many of the water stations along the course—one roughly every mile—taking cups of Gatorade. At mile 14, still going strong, she high-fived the women from Wellesley College who lined the course to cheer on the runners. Eight miles later, at Cleveland Circle, her legs suddenly went rubbery. She told a friend that she felt thirsty. Then she passed out.

She was rushed to Brigham and Women's Hospital, where blood tests revealed a dangerously low concentration of sodium in her blood, just 113 millimoles per liter; normal is more like 140 mmol/l. This is called *hyponatremia*—technically, low blood salinity, which sounds like no big deal but is a very serious and potentially fatal condition.

We need to maintain a certain level of saltiness in our bodies for things like cell membranes and neurotransmitters to work properly. When our sodium level gets too high, as when we are running and sweating out fluid, we feel the urge to drink; if our salinity goes too low, we expel excess water by urinating. But if those homeostatic mechanisms somehow fail, and water accumulates where it shouldn't, bad things can happen, as they did to Cynthia Lucero. Hospital tests revealed that her brain had swelled up because of all the excess liquid in its cells, a deadly condition called exercise-associated hyponatremic encephalopathy (EAHE). She never woke up.

Two days later, doctors removed her from life support. She was twenty-eight.

Cynthia Lucero was only the second person to have died in the one-hundred-five-year history of the Boston Marathon, but it was the cause of her death that shocked the running community. The medical examiner later concluded that she had died at least in part from drinking too much fluid during the race.

Yet she had done precisely what the science at the time had advised: Hydration guidelines published by the American College of Sports Medicine (ACSM) in 1996, and echoed in marketing campaigns by Gatorade and other sports-drink companies, urged endurance athletes to drink enough water or sports drink to replace *all* the fluid that they might lose to sweating—or, barring that, to "*consume the maximal amount that can be tolerated* without gastrointestinal discomfort" [emphasis added].[4]

She had done just that, and it had killed her.

When Tim Noakes heard about her case, his jaw dropped in recognition. Noakes is the South African sports scientist who authored the Central Governor theory of performance, which we talked about in the previous chapter. He knew exactly what had happened, because he had encountered a similar case of exercise-associated hyponatremic encephalopathy (EAHE), the brain swelling that killed Lucero, back in 1981, when a forty-six-year-old female runner had died at the popular 55-mile Comrades ultramarathon in Cape Town.

Lucero's death proved to be only the most visible tip of a gigantic iceberg of unseen, unacknowledged cases of hyponatremia in runners and other endurance athletes. Relatively few athletes had died, but it turned out that hyponatremia was far more common than anyone suspected. A study done at the 2002 Boston Marathon, the same race where Lucero had died, discovered that some 13 percent of a sample group of about five hundred recreational runners—one in eight—had finished the race with some degree of hyponatremia (defined as blood sodium below 135 mmol/l). Only a handful had clinically critical hyponatremia (<120 mmol/l), but clearly, many runners were heading in that direction.[5]

Similar findings had been observed at other events, notably an Ironman triathlon in New Zealand where 18 percent of a sample of finishers were reportedly hyponatremic. Many athletes, it appeared, were at least flirting with hyponatremia. And slower-running women, like Lucero, were at special risk, because they spent more time on the course and were more likely to stop at every water station. Yet almost nobody had noticed.

Making it even scarier is the fact that key symptoms of hyponatremia overlap with those of dehydration, including upset stomach, erratic or unhinged behavior, and (bizarrely) feelings of extreme thirst. The only way to distinguish hyponatremia from dehydration with certainty is via a blood sodium test, which is not easy to do in the middle of a marathon or mass bike ride. As a result, many cases of hyponatremia went unrecognized, or were worsened by well-meaning race medics who give such patients the exact opposite of what they need: still *more* fluids.

Noakes zeroed in on the then-current hydration guidelines that advised athletes to drink as much fluid as they could handle.* They appeared to be well-supported by published research, drawing on

---

* I was racing mountain bikes "avidly" during this same period, and I well remember how my teammates and I would aggressively guzzle fluids before, during, and after competition. (I also well remember the bloated, queasy feeling that often accompanied these binges, and the occasional midrace vomiting, which we blamed on overexertion rather than overdrinking.) We had not read any scientific guidelines, but the athlete's mantra at the time was *Hydrate or Die*. Little did we suspect that it was even possible to hydrate *and* die.

multiple studies going back to the 1970s that confirmed the importance of hydration for athletes.

He wondered: What if we had the science of hydration completely wrong?

Water is the elixir of life. Our bodies are made up of about 50 to 60 percent water; more in young children, slightly less in elderly people. That fluid is distributed all over our bodies—not only in our blood, but inside our cells, outside our cells, in our tissues and our organs. Our muscles and kidneys are comprised of around 79 percent water by weight, and even bone is nearly one-third water. A 180-pound male will be carrying close to 100 pounds of water in his body, give or take. "We're big bags of water," says Minson.

Humans can survive for weeks without food, but a person who is deprived of water will die within days. This is the price we pay for the miraculous sweat-cooling system that evolution gave us. We are sweaty apes, but also thirsty apes, and sooner or later we must replace the water we lose. One study using deuterium-labeled water found that under hot conditions, the fluid that we drink can appear as sweat on our skin in as little as three minutes.[6]

"Water is the indispensable, limiting factor for man in the desert," a team of US Army physiologists declared, after spending months studying soldiers training in the Mojave Desert during World War II. In warfare, they concluded, water ranked in importance "second only to ammunition."[7]

Until the 1970s, however, athletes were still being warned (as they had been for decades) to avoid consuming fluids altogether while competing—especially in an event as long as the marathon, because it was believed that the water would slosh around in the athletes' stomachs and make them sick. Cycling was even worse: Riders in the Tour de France would stop at cafes and guzzle beer, wine, coffee, and even brandy—anything *but* water—before hopping back on their bikes. As late as 1980, half of the 5,400-plus finishers of the Melbourne marathon had drunk no water at all during the race.[8]

Now it has gone completely the other way. At major marathons and other events, runners are loaded down by their "hydration systems," lugging water bottles, water belts, and water packs filled with specialized sports drinks of every description—even though, in most races, they will likely encounter an aid station every mile or two, with volunteers shoving cups of water and sports drink at them. The hydration mania is not limited to athletes: Online health influencers urge people to drink at least eight 8-ounce glasses of water per day, which works out to half a gallon. And it is impossible to board an airplane now without getting clonked by someone's 40-ounce stainless-steel water bottle.

What happened?

In a word: Gatorade.

Many people are familiar with the 1960s origin story of Gatorade, the granddaddy of sports drinks. As the legend goes, the University of Florida football team had been losing games in hot weather, so the coach asked Dr. Robert Cade, a kidney specialist at the UF medical school, to create a drink that might help the players perform better. Cade mixed up a concoction of water, sugar, salt, and potassium that tasted (by some accounts) like kitchen cleanser until his wife suggested adding lemon juice. After the Florida Gators won the Orange Bowl in 1967, crediting their success to the special yellow potion, it became known as "Gatorade."

Prior to that, nobody had really thought much about hydration and athletic performance. But Gatorade proved to be a hit. Cade formed a company, and sales quickly rocketed into the millions. Early Gatorade advertisements claimed that the magic potion was absorbed twelve times faster than water. Gatorade was eventually bought by Van Kamp and then Quaker Oats, before ultimately being absorbed into the massive global beverage empire of PepsiCo.*

As Noakes dug into the research around hydration, he spotted a glaring issue: Many if not most of the key studies underlying the hydration guidelines had been funded by Gatorade itself, posing an obvious conflict of interest. In 1985, Gatorade had established the Gatorade Sports Science Institute, with a large lab in Illinois (since moved to Westchester

---

* The whole story is told in Darren Rovell's 2005 book *First in Thirst*.

County, New York), specifically to generate research into hydration and other sports-performance-related topics. Meanwhile, the company also funded several leading sports-science organizations, including the American College of Sports Medicine (ACSM), which promulgates hydration guidelines for athletes. This blurred the line between science and marketing, Noakes felt.

Outraged, he prosecuted his case against Gatorade, and what he viewed as flawed science around hydration, in a heavily footnoted, extremely dense 400-page Jeremiad of a book called *Waterlogged: The Serious Problem of Overhydration in Endurance Sports*. In Noakes's view, the hydration guidelines were not only wrong, but also dangerous, potentially leading athletes—especially less fit, slower recreational athletes like Cynthia Lucero—down the road to hyponatremia.

He was flying in the face of the scientific consensus. Decades' worth of studies had consistently found that dehydration impairs athletic performance, and possibly puts athletes at risk for heatstroke. In one especially influential paper from the early 1990s, two highly respected sports scientists had demonstrated that as study subjects grew progressively more dehydrated, their athletic performance declined in lockstep—while their heart rate and core body temperature steadily and dangerously increased.[9]

Indeed, the conventional wisdom held that even mild to moderate dehydration, if allowed to continue, would *inevitably* lead to heatstroke in hot conditions. Yet none of the subjects in any of these studies had actually suffered from heatstroke due to dehydration. For good reason—research-safety protocols required scientists to halt experiments when subjects' core temperatures reached the potentially danger zone of 104 degrees F (40 degrees C) or above. Actual heatstroke is therefore almost impossible to study in a lab.

But back in World War II, in those desert studies I mentioned earlier, the military scientists had been able to push the soldiers to their very limits. They had put the men through a grueling series of experiments that they called "dehydration hikes," desert marches lasting as long as nineteen hours. On some hikes they wore full uniform; on others, light clothing; and on still others, the men hiked nude (which turns out to be a bad idea, in the desert). Sometimes the soldiers were permitted to drink after thirty minutes, sometimes after two hours, and sometimes not at all, even on hikes of more than ten miles. The scientists hiked along, too, observing the soldiers as they slowly fell apart.

The more dehydrated they became, the more the men suffered. Thirst kicked in at about the point where they had lost 2 percent of their body weight in water. At 3 to 4 percent, they felt "vague discomfort" and began to complain. Five to 6 percent brought apathy, lagging pace, and stumbling. Seven to 8 percent, indistinct speech. Delirium set in at 12 percent, followed by inability to swallow at 14 percent, and it got worse from there: shriveled skin, sunken eyes, dim vision, painful urination, even deafness.

On one especially long trek, the men who were deprived of water began going slightly crazy, understandably. Some grew "peevish and intractable," the researchers later wrote, while others fell silent and unresponsive, marching like zombies. One desperate individual cut open a cactus and chewed it like a madman. As the situation deteriorated, some men in the water-deprived group attacked a hospital truck that was following the marchers in case of emergency.

On the other hand, when the men were provided sufficient water (and encouraged to drink it), they could hike almost indefinitely, even on days that were as hot as 107 degrees F in the shade—after a week or so of heat acclimatization training.[10] Even if they became moderately dehydrated while hiking, they would recover once they had rested and replenished their body water. But without water, it was a different story. Heatstroke was not the problem on these marches; at a certain point, the men would simply stop walking, unable to continue—their thirst having rendered them "incapable of even mild physical effort," the researchers wrote. Running was a different story, however, and the men could get into heat trouble fairly quickly if they tried to run in the desert heat, the scientists observed.

More recent data from the US military, which continues to research heat and hydration issues intensively, further questions a link between dehydration and heatstroke. The US Army had also recognized its own potential issue with hyponatremia, as far back as the late 1990s, and had adjusted its own hydration guidelines to prevent recruits from overdrinking.

As US Army Col. David DeGroot, a heat researcher and director of the Heat Center at Martin Army Community Hospital in Fort Benning, Georgia, has observed, a substantial portion of exertional heatstroke cases occur in soldiers who are in fact well hydrated. Dehydration is more commonly associated with cases of heat *exhaustion,* a much less serious condition than heatstroke.*

So while dehydration is *a* risk factor for heat illness, notes DeGroot, it is not the only or even the most important causal factor for heat*stroke*: "Excessive focus on hydration may result in overlooking other risk factors," he said.[11]

The next question is more complicated: Does mild dehydration impair athletic performance?

The scientific consensus insisted that it does; even a 2 percent loss of bodyweight, some studies suggested, is enough to slow athletes down significantly.

But if that was so, Noakes wondered, then why were athletes winning major events having lost 5, 6, even 7 percent of their bodyweight? "The interesting thing is that in competition, often the most dehydrated athletes win the race," Noakes says.

The legendary Ethiopian runner Haile Gebreselassie lost nearly 10 percent of his body weight en route to winning the Berlin Marathon in 2008, setting a world record in the process. He barely drank anything throughout the entire race, yet his performance didn't seem to be impaired. But according to the accepted science of the time, he should have been at death's door, not the winner's podium.

Gebreselassie was one of the world's greatest runners, so perhaps he was an outlier. But this same phenomenon had been observed often in amateur athletes as well. One large study of Ironman triathletes found that they often finished their multihour races with hydration deficits of

---

* The main differences: Heat exhaustion symptoms include dizziness and nausea, while exertional heatstroke occurs at core body temperatures greater than 104 degrees F and involves extremely severe neurological symptoms where the victim is likely to be confused or even seemingly deranged. Both are considered heat illnesses, but heatstroke is obviously much more serious and potentially life-threatening. Some physiologists believe that heatstroke is not caused by extreme temperature alone, but also by the effect of the sun, which appears to induce the neurological symptoms associated with true heatstroke.

4 percent or more, some losing 11 or 12 percent of bodyweight.[12] Real-world results like this appeared to contradict decades' worth of laboratory studies finding that athletes' performance fell off a cliff at 2 or 3 percent water loss.

Then a Canadian researcher named Paul Laursen, working at an obscure university in Australia, noticed a problem with many of those laboratory-based hydration studies. In virtually every single one, the subjects had *known* that they were dehydrated, typically because they had been forbidden to drink for several hours, or they had exercised in the heat; this was how they had become dehydrated in the first place. They were already thirsty. But this also meant that the studies were not truly blinded.

"The group that's getting the hydration, they're probably feeling like they're on top of their game," Laursen told me. "Whereas in the [dehydrated] condition, potentially they might feel stressed." Stressed athletes tend to be slower athletes; their Central Governor is stingier with the throttle. Other studies had found that athletes who were dehydrated had higher core temperatures, elevated heart rates, and felt hotter from the very beginning of their workout.

Laursen tested this idea by means of a clever experiment. First, he had his study subjects get slightly dehydrated, by 3 percent of their body weight, by exercising and sweating, or not drinking for several hours. Then he replaced some of that liquid intravenously, but without letting the subjects know how much fluid they were getting. He then had each athlete ride a 40-minute time trial in the lab, three different times, at three different hydration states: fully hydrated, 2 percent dehydrated, and 3 percent or more dehydrated by body weight. There was no difference in their times. Their hydration status did not affect their performance at all.[13]

One wonders, however, how the athletes in Laursen's study would have fared over the course of a three-hour run or ride, as opposed to 40 minutes on a stationary bike in a lab. The riders' body temperature had already begun to creep upward by the end of the 40 minutes, when they began the time trial at 3 percent dehydration. What if they had continued, and their 3 percent dehydration progressed to 4 or 5 percent, or more? At that point, things start to get miserable: heart rate speeds up, thirst intensifies, stress and irritability set in, focus slips. At extreme levels of dehydration, runners may experience hypovolemia, sudden low blood pressure from reduced plasma volume—a serious condition that can lead to unconsciousness, organ damage, even death.

Yet at the same time, it is physically almost impossible for athletes to replace *all* the fluid they sweat out in real time, especially in hot conditions. Elite marathoners are simply running too fast to stop and guzzle water. But even less trained athletes struggle to replace their fluids completely. For example, my "sweat test" at the Korey Stringer Institute revealed that I sweat out about a liter of water per hour in Hotter'N Hellish conditions—about average for a 175-pound man. Some athletes sweat far more than that; Ryan Swoboda lost four liters per hour. There is no way any of us can replace that fluid. A typical man's stomach can absorb no more than 0.8 liters of liquid per hour, and much less during intense exercise in the heat. The math simply does not work out.

The bottom line is that a little bit of dehydration is almost inevitable, even more so in athletes who sweat excessively. Does this matter to performance?

Only to athletes who are pushing their limits in hot conditions, says Douglas Casa of KSI, who tangled often with Noakes over hydration issues: "The caveat is that this *only* matters during intense exercise in the heat," he says. "Hydration does not matter [as much to performance] if it's cool, or if it's not intense. That relationship does not hold."

The questions raised by Noakes and Laursen and others—and the shocking death of Cynthia Lucero—ultimately led to a softening of the old, dogmatic approach to hydration in sports. Since 2002, hydration guidelines from ACSM, USA Track & Field, and other bodies have been relaxed considerably, and athletes are now encouraged to pay more attention to their actual fluid losses from sweating, rather than just blindly guzzling fluids throughout their events.*

Most importantly, Noakes had successfully raised awareness of the deadly danger of hyponatremia and overdrinking among athletes. "He

---

* The closest thing to a blanket recommendation comes from the ACSM, which suggests that athletes consume between 13 and 27 ounces per hour (400–800ml/hr.) while training or competing, depending on conditions and the individual's sweat rate—which is a pretty wide range.

saved lives," says Casa. But among the public, the message has yet to get through—as the actress and model Brooke Shields learned very dramatically.

One afternoon in September 2024, Shields was waiting for an Uber outside her home in New York City, on her way to perform at an Upper East Side jazz café, when she suddenly felt weak and staggered into a neighboring restaurant. Before she could explain herself, she turned blue and collapsed, frothing at the mouth, right in front of the sommelier.

In the hospital, doctors explained that she had suffered a grand mal seizure caused by overhydration. "I had had too much water," she told *Glamour* magazine a few days later. "I flooded my system, and I drowned myself."[14]

Just like Cynthia Lucero, Shields had been doing what she thought was the healthy thing: drinking lots of water. And she had nearly killed herself. She was lucky compared to Ashley Miller Summers, a thirty-five-year-old mother of two from Indiana who died of what doctors called "water toxicity," after guzzling four bottles of water following a hot Fourth of July boating outing with her family.[15] In another notorious case, from 2007, a twenty-eight-year-old California mother died after a water-chugging contest sponsored by a radio station. Her husband later won a $16.5 million jury verdict against the station,[16] in part because multiple people, including a nurse, had called the station during the contest, warning them in real time of the dangers of what they were doing.*

These cases are outliers; full-on clinical hyponatremia is extremely rare. It appears that at least the *necessary* condition for hyponatremia, however, barring heart failure or some other medical condition, is drinking way too much water, way too quickly. Which an awful lot of people seem to be doing, whether they realize it or not.

While hyponatremia and water toxicity are at least recognized in the world of endurance sports, they remain virtually unknown among the general population. On social media, on health podcasts, and in mainstream media, we are bombarded by messages encouraging us to be

---

* In the contest, called "Hold Your Wee for a Wii," the contestant who drank the most water without urinating would win a then-cool Wii video game controller (this was 2007).

vigilant about our hydration; carrying a water bottle has become a kind of social signal that you are healthy, athletic, and attractive.

The messaging is especially relentless on women. Beyoncé has said that she drinks a gallon of water a day, with lemon. Ex-*Friends* stars Jennifer Aniston and Courteney Cox agree that drinking a lot of water is their beauty secret, keeping their skin smooth and their weight optimal. Not to be outdone, celebrities including Gwyneth Paltrow, Victoria Beckham, and Drew Barrymore swear by their crystal-equipped water bottles that are said to "charge the water with vibrational energy."[17]

I was tempted to bring one to the Hotter'N Hell.

It doesn't take much digging to find the influence of the beverage industry behind this craze for hydration, often in industry-sponsored studies suggesting that low levels of dehydration can be harmful not only for athletes, but in everyday life. In one of the most blatant examples, a 2015 study from the University of Loughborough in the UK found that drivers who became only slightly dehydrated became more dangerous behind the wheel—almost as bad as drunk drivers, the researchers claimed.[18]

This study generated the predictable scary headlines, but few news stories mentioned that it was funded by an organization called the European Hydration Institute, which in turn was sponsored by Coca-Cola, owner of Powerade, Gatorade's biggest rival. A few months after the study appeared, Coca-Cola launched an advertising campaign urging British drivers to "stay hydrated through the day," citing the study as justification. The goal of the ad campaign, a partnership with Shell, was to boost drink sales at highway service stations according to an investigation by *The Times* of London. Campaigns like this have succeeded beyond expectations: Bottled water, sometimes just purified tap water, now outsells any other beverage category.

Usually, drinking an extra Coke or Powerade or Dasani does no harm; even guzzling water all day is fine for most people, most of the time. The worst outcome is usually a few (or many) more trips to the restroom. Also, there is some evidence that drinking more fluids may reduce the incidence of kidney stones and bladder cancers, as well as of migraine headaches. This is good. But assaulting your body with vast quantities of liquid, all at once, is another matter.

That is the intent of the most extreme hydration totem yet, a one-gallon monster water bottle, popular on TikTok and sold on Amazon (endorsed, in one case, by a Kardashian), marked with motivational slogans and timestamps, from "HYDRATE YOURSELF" at 9 a.m. to "REMEMBER YOUR GOAL" at 11 a.m. to "KEEP CHUGGING" at 1 p.m. to "DON'T GIVE UP" at 5 p.m.

Actually, *do* give up, if you feel like it. I'll explain why.

The fluid balance in our bodies is regulated by a series of sensors that detect the "osmolality" or saltiness of our blood; imagine these sensors as essentially tasting our blood, to determine if it is too salty, or not salty enough. "If I'm sweating copiously, my blood salinity goes up because I've lost fluid," Huggins explains. "My osmoreceptors tell me, your [sodium] concentration in your blood is too high, and you need to replace that fluid."*

Clinical hyponatremia is almost always caused by a breakdown in this water-balance system, for various reasons. It could be caused by kidney disease, heart failure, and certain medications, such as ibuprofen and SSRI antidepressants. Regardless of the cause, hyponatremia typically occurs when the body decides to retain excess water when it should be expelling it. If that excess water seeps into the person's brain cells,[19] it could lead to them passing out in the foyer of a restaurant, or even dying in their garage. Or at the Boston Marathon, as happened to Cynthia Lucero.

Yet social media is full of health influencers pushing extreme hydration. Its most sciencey-seeming form is the widely repeated recommendation to drink eight glasses of water per day. This notion, also known as 8-by-8, has been around so long that it seems science-based, but has little basis in evidence—and was thoroughly debunked nearly a quarter century ago. In a 2002 paper in the *Journal of Physiology*, Dartmouth nephrology (kidney) professor Heinz Valtin searched for scientific

---

* The signal for this comes via a hormone called vasopressin, also known as ADH or antidiuretic hormone, which (as its name implies) tells the body to retain water; other signals prompt the person to drink, due to thirst. When we go to sleep, our vasopressin rises, to stop us from having to urinate during the night. Conversely, if our osmolality drops too low, vasopressin also decreases, and we pee. So if we drink alcohol before bedtime, that also suppresses vasopressin or ADH, which is why we often wake up at odd hours after a night of drinking.

support for 8-by-8 and found none; there were no clinical trials and no other data suggesting that people need this much water.*

"On the contrary," Valtin wrote, "there are publications that state the opposite."[20]

Nearly twenty-five years later, 8-by-8 persists, a kind of zombie myth. Where did it come from? It first appeared in an obscure 1921 study where the researcher measured his *own* daily water intake and extrapolated from that to declare that therefore *everyone* needed to drink about two liters per day. This might seem weedsy, but it illustrates perfectly why these kinds of blanket recommendation ("8 glasses a day") rarely make sense, in any area of biology. People vary tremendously, in every aspect of our lives—what time we like to go to bed, what kind of food we like to eat, our athletic ability, and so on.

Hydration is no different. One large global study of water turnover found that some individuals needed up to five times more water per day (from all sources, including food) than others; the variation depended on body size, biological sex, activity level, socioeconomic status, the altitude and even the latitude at which they lived. [21]

My point is, one-size-fits-all dogmas are useless and possibly dangerous, because they could lead some people to drink far more than they need or can tolerate, while others might not get enough fluid. For example, Paul Laursen says that he often goes on multihour bike rides in the mountains near his home in British Columbia with just one water bottle, which he rarely finishes. If I attempted that same ride, with less than two 22-ounce water bottles on board, I would be miserable. Rather than heeding prescriptive advice from strangers or succumbing to social-media "challenges," most of us are better off paying attention to signals from our own bodies—especially the built-in hydration gauge and drinking prompt called *thirst*.

So how much water should we drink when we exercise in the heat?

The current hydration guidelines put out by the various sports

---

* The reason Valtin suspected 8-by-8 was misguided was because of a weird bit of homework he assigned his students every year: For one 24-hour period, they had to capture every drop of their urine in a plastic jug and then measure the total. Over the years, he calculated that his students consumed more like 1.2 liters of fluid per day, on average, or about 40 ounces—only five 8-ounce glasses, a more reasonable amount.

medicine and athletic trainers' associations are less dogmatic and one-size-fits-all than they were decades ago—but if anything, they have gotten far too complex and opaque even to summarize here. (If you're interested, check the references in the endnotes.[22]) At the end of the day, the best advice is probably: Don't overdrink, but also pay attention to how thirsty you are, how much you are sweating, and how hot it is. We lose water all day, even by breathing, so match your intake to your needs.

For events lasting an hour or less, according to research by Stephen Cheuvront and Robert Kennefick, two veteran heat researchers with the US Army, hydration doesn't even matter all that much; it isn't necessary to bring a water bottle into a yoga class or an exercise class, as long as you start out well hydrated.[23] (Obvious as it may seem, this is the Achilles' heel of many of those laboratory studies we've been talking about, where subjects started out *already* dehydrated. No wonder they struggled from the get-go.)

For longer events, up to about four hours, the researchers say, one should aim to replace roughly half of one's fluid losses in real time, although that could vary depending on body size and gender; a 130-pound female can become dehydrated (by percentage) more quickly than a larger male in some situations.[24] Also, consider the conditions: If it's hot, obviously, you will need to drink more; if it's cool, not so much. (But be careful to drink something.) If you are working hard, riding, or running fast in the heat, you will obviously sweat more than you would at a more moderate pace.

Most importantly, know thyself—how much do you sweat? What kinds of drinks satisfy your thirst and keep you going? What drinks taste good to you? How often will you be able to refill your bottles or other hydration systems, along the way?

Which is why I was glad I had done the Sweat Test. It was useful to know in advance exactly how much I sweated while exercising in the heat, and how salty that sweat was; I had never even really thought about that before. Even in my "avid cyclist" days, I would guzzle sports drink blindly—because that's what the dogma of the time said to do. If anything, I now realized, I needed to be careful not to drink too *much*.

As far as electrolytes, the Sweat Test revealed that I lost about 800 milligrams of sodium with that sweat, which is also typical. In a shorter event, like an hour-long run or exercise class, most people don't need to worry about electrolytes; most people living in Europe, Australia, or North America will get quite enough salt and potassium from their diet alone, Huggins told me, thanks to that popular athletic superfood known as "French fries." Over the course of the Hotter'N Hell, however, that wouldn't cut it; I would need to pack some type of electrolyte drink or tablets. (I picked fruit-flavored salty chews called "Bolt.")

And while all this knowledge came from fancy sports-science testing at KSI, it is easily accessible to anyone with a scale and a few extra bucks.

Calculating your sweat rate is simple: Before working out or going for a run, weigh yourself, making sure you are holding any water bottles or hydration systems you plan to use during the workout. But no clothes: "Get naked," says Huggy Bear. Then, after the run or ride (or golf game, or whatever), weigh yourself again—also naked, and holding those same water bottles. The difference in weight, divided by time, represents your overall sweat rate, roughly. (Be sure to account for refilling those water bottles, if you did that, and/or going to the bathroom.)

If you didn't already know how sweaty you are, now you do. Make before-and-after weighing part of your workout routine. If you find yourself gaining weight during an athletic event or a training session, you are likely drinking too much. If you are losing significant weight, try drinking more, if you can. It's all about keeping in balance.

Another useful gauge: urine color. If your pee is pale yellow, like a dry Sauvignon Blanc, you're good; clear like Evian water may indicate overhydration, while dark like Sam Adams beer suggests dehydration or some more serious health issue.

While I trained for the Hotter'N Hell, I tracked my sweat losses and hydration status—and my electrolyte needs—via an ingenious little device called the Nix, which consists of a sensor patch that sticks to your upper arm. Invented by a recreational marathoner and venture capitalist named Meridith Cass, after a scary brush with hyponatremia, the Nix measures fluid loss via sweating and senses the electrolyte content (saltiness) of that sweat, *while you are exercising*. It allows you to monitor your hydration status in real time, and prompts you to drink at regular intervals, even suggesting what kind of sports drink best matches your sweat profile. It helped me pace my drinking, so I drank neither too little nor too much.

Gatorade makes an even simpler device, consisting of a patch that you stick on your forearm before you work out. As you sweat, the patch changes colors; you then scan it with your phone to determine how much you sweated during your workout, and what type of (Gatorade-branded) hydration product you should be using.

As for *what* to drink: Sports drink manufacturers all claim that their unique mix (of liquid, electrolytes, possibly sugar or other carbohydrates, and/or protein) is optimal for hydration, and certainly better than water. But research comparing the hydration potential of different beverages—basically, how much of the fluid is retained, as opposed to simply being urinated out—calls these claims into question. (Not surprising by now.)*

Carefully controlled studies have found that most sports drinks perform little better than plain old water, and are slightly *less* effective at rehydration than oral-rehydration solutions such as Pedialyte, which are designed to treat serious dehydration due to illness. Even Coca-Cola compares well to the big-brand sports drinks, weirdly, which is one reason why exhausted Tour de France riders will gladly accept an ice-cold can during a hot mountain stage. And more than one study has found that one of the most effective beverages of all—meaning, the one that best replenishes lost fluid—is milk.

Not that anyone would want to drink milk while running a marathon, but that's what the studies found.[25]

Also unexpected: Caffeinated beverages seem to be just as effective as non-caffeinated drinks in terms of rehydration potential,[26] and one clinical trial from Australia found that drinking coffee an hour before competition might even improve exercise performance in the heat.[27] This is counterintuitive, but yet another study found that drinking a hot beverage like hot coffee or tea may even help cool the body, by raising core body temperature and thus causing one to sweat more and sooner.[28]

---

* My *Outlive* coauthor Peter Attia believes that some sports drinks may actually *increase* dehydration, net-net, because of the way in which their sugars are metabolized in the gut. https://peterattiamd.com/ama33/.

I can hear you silently wondering, *What about beer?*

Good news: Low-alcohol or lager beers perform surprisingly well in hydration studies (huzzah!)—comparably to sports drinks, in one study anyway, and slightly better than water, which is a pleasant surprise. High-alcohol beer proves to be rather less hydrating, but perhaps satisfying in other ways.[29] (Avoid IPAs, which dry out your mouth like a Mojave Desert trail.) One study from Australia found that low-alcohol beer (2.3%) with a little bit of added salt proved to be significantly more effective than plain full-strength beer for rehydration.[30]

Sadly, such a concoction was not likely to be available at the rest stops at the Hotter'N Hell Hundred. Also, it would probably be a bad idea to drink *only* beer, or milk for that matter, in hundred-degree F heat. I would need to stick to my original "hydration plan"—which turned out to be easier said than done.

7

HOTTER'N HELL

> Manage the heat, let the meat cook,
> and you'll get fantastic results.
> —GUY FIERI

Two days before the Hotter'N Hell Hundred, in late August of 2023, I flew to Dallas and drove two hours northwest to Wichita Falls. I had planned to give myself an extra day to acclimate to Texas, but it soon became clear that one day would not be enough. When I arrived at 9 p.m., the temperature gauge in my rental car said it was still 95 degrees F.

Back in mid-June, a "heat dome" had parked itself over most of Texas, and it hadn't budged. A heat dome is a meteorological phenomenon where a stable area of high-pressure air forms and then just hangs out, trapping the heat and blocking any storms or cool fronts from moving through and breaking things up. On the weather map, it looked like most of the state had been swallowed by an enormous red-orange blob.

Heat waves come and go, but heat domes like to stick around. This one was a doozy. From mid-July up until the day I flew to Dallas, the high temperature in Wichita Falls surpassed 100 degrees F on all but two days.* Meanwhile, life went on. The following night, Friday, the

---

* For two blessed days in mid-August, the heat broke, with highs of only about 90 degrees F. But the intense heat came right back and it stayed hot right up through the weekend of the ride.

two local high school football teams would be playing in front of more than fourteen thousand fans.

All the hotels were full, so I had found one of the last rentals, a garage that a nice young couple had converted into a tiny guesthouse. It had a DIY split-system air conditioner that labored mightily all night long, blasting frigid air in my face and then switching off to rest. By dawn, I was damp and sticky with dread.

There was, not surprisingly, a heat advisory in effect for Wichita Falls and surrounding areas—as there had been nearly every day since June. The National Weather Service instructed residents to "drink plenty of fluids, stay in an air-conditioned room, [and] stay out of the sun." We were also supposed to "check up on relatives and neighbors," to be sure they weren't doing anything dumb, like riding their bikes 100 miles to nowhere in the middle of the day.

I woke up hungry and went out for a breakfast of greasy but delicious chilaquiles. (There was no official warning against that, but maybe there should have been.) Then I went back to the guesthouse to digest and contemplate my poor life choices. My longest training ride had been 47 miles, less than half the distance of the Hotter'N Hell, and that was in Michigan, where it was about 20 degrees F cooler than Wichita Falls. Nine words echoed in my head: *You'll never make it, because you're not from there.*

After worrying away the morning, I decided to go for a walk around the neighborhood—a fancy part of town called Country Club, with quiet streets curving around stately old homes. There was nobody outside besides some landscaping workers and a USPS driver who kept leapfrogging me in his van, giving me a funny look each time. The heavy, humid air seemed to smother all sound. In front of one house, three workmen sat on the curb eating lunch, taking a break from repaving someone's driveway. Even they looked at me like I was crazy.

Most of the homes looked to have been built during the 1920s heyday of Wichita Falls—stately Tudor-style and Italianate mansions, a miniature chateau or two, and what looked like a replica of the White House. Built before widespread air-conditioning, the homes all had thick walls and high windows, shaded by tall trees. Yet according to Zillow, they were bargain-priced, for mansions. You just had to be able to handle the heat.

I wasn't sure I could. Just walking around was exhausting; I couldn't imagine riding a bike in this. But then something happened. After about 20 minutes, I rounded a corner near the actual Country Club, and a gentle breeze fluttered my shirt. That evoked an unfamiliar sensation: I felt cool. Of course—I was sweating. My shoulders and chest were damp. But the light wind lifted away some sweat and with it, a bunch of body heat. I shivered with delight. Evolution had done its job; I was "thermoregulating," just like our ancient ancestors.

I took a deep breath and let it all out. *You can do this.*

It was pitch-dark the next morning when we riders began assembling in the back streets of Wichita Falls, pulling bikes out of cars and adjusting shoes and helmets before making our way to Scott Avenue, the main drag.* It was already 77 degrees Fahrenheit, according to the bank thermometers, and the air felt thick and moist.

The streets slowly filled with cyclists until we stood shoulder to shoulder for blocks in both directions. At sunrise, tinny speakers blared the "Star-Spangled Banner" as four vintage military aircraft from nearby Sheppard Air Force Base rumbled overhead in a slow-motion flyover. Even they seemed to struggle through the soupy air. Then a cannon boomed, and all nine thousand of us surged forward as one, as the air filled with Texan whoops and hollers. It was thrilling, like a concert or a demonstration or a parade, thousands of humans united in one common purpose.

The first hour passed almost pleasantly as we flowed west down the four-lane highway out of town, a river of whizzing, whirring wheels and gears. Locals lined the road in folding chairs, simultaneously cheering us on and, like NASCAR fans, half-hoping to witness some gnarly wrecks. There were a few tangle-ups, a few more flat tires, and some outbreaks of yelling, but with the sun still low in the east, it felt like

---

* Scott Avenue was named for John A. Scott, who had owned the land where Wichita Falls was built. According to legend, he won it in a poker game in 1837. (Probably not true, but a good story anyway.)

no big deal. Riders sped up, jockeying for position as our competitive juices began flowing.

And then, at a little after 9 a.m., as if someone had flipped a switch, it got hot. The road tilted upward in a gentle climb, and we suddenly became aware of the weight of the sun on our shoulders and backs. The dense river of cyclists had stretched out into a long trickle reaching all the way to the horizon as we pedaled along this Farm-to-Market road in the middle of, let's just say it, nowhere. Nobody talked.

The rest stops spaced every 10 or 12 miles were a welcome relief, big shady tents staffed by friendly local volunteers. David Martin, my coach from KSI, had prescribed a detailed nutrition and hydration plan, indicating what I was to eat and drink at regular intervals during the ride. "Just stick to your plan, and you've got this!" he had said on the phone the previous day, trying to quell my panic attack. My jersey pockets and handlebar bag were stuffed with packets of the requisite sports drink mixes and athletic "food"—gooey liquid gels and gummy blocks that are packed with vitamins, electrolytes, and especially carbohydrates, pure fuel for high-performing endurance athletes like me.

All good, in theory. But in real life, they proved unappealing next to the barbecued hot dogs and smoked sausages on offer at the rest stops. In my defense, those tasted like they had plenty of protein, fat, and electrolytes in them. Just to be safe, I washed down this high-performance athletic fuel with a briny concoction called Pickle Juice that was incredibly salty and weirdly satisfying, especially when chilled. It would have tasted amazing in a martini, but none of the rest stations appeared to have gin on hand.

The Hotter'N Hell is legendary in the history of endurance sports—and of hydration, in particular—as the birthplace of the CamelBak. For the 1989 edition of the ride, a clever EMT from Dallas named Michael Eidson borrowed an IV bag from work, filled it with water, and slipped it into a white tube sock to keep it cool. He put the whole thing in the back pocket of his jersey, slung the clear plastic tube from the IV over his shoulder, and sipped water as he rode along. His makeshift device revolutionized hydration. Today hydration packs by CamelBak and its many imitators are used not only by millions of athletes, but by military and law-enforcement groups around the world.

The genius of the CamelBak is that it allows one to drink as much as one desires, without needing to stop or take one's hands off the handlebars. This has pluses and minuses, as we discussed in the last chapter, but halfway through the Hotter'N Hell, I could see the appeal. It was so hot that it felt like we were riding through an air fryer, being slowly roasted by the sun and the wind. There was barely any shade, except at the rest stops, where fresh-faced high schoolers refilled our bottles with water and precious ice. Every half-mile or so, I would take a refreshing sip, then pour a little bit across my neck and shoulders. That brought fleeting relief. Then it was back into the air fryer.

My thirst was insatiable. No matter how much Pickle Juice and ice water I drank, it felt like it was never enough. I was always thirsty, counting the miles until the next rest stop. Yet, at the same time, I began to feel liquid sloshing around in my belly. My mouth was parched but my body felt bloated. That spiraled into: What if I drank too *much* water? Am I having brain fog? Is that a symptom of dehydration? Or hyponatremia?

It was turning out to be one of the hottest Hotter'N Hells in years, as the temperature climbed past 100 degrees F and kept right on going. It was stunningly, paralyzingly hot. Back on the road, heat blazed up at us from the blacktop, while the Heat Dome roasted us from above. A rumor spread that the ride would have to be cut short because of the extreme heat. Even at the Hotter'N Hell, that had rarely happened.

At each successive rest stop, our fellow riders looked just a little bit sweatier, paler, and more tired. So did I. The medical cots were now mostly occupied, by riders sitting and staring into space, some of them hooked up to IV bags. Nobody seemed all that eager to get back on their bikes. With good reason: It stopped being "fun" many miles ago. The pell-mell rush of the early hours had been replaced by a *Hunger Games*–like struggle to survive.

The Turtle Boys saved me.

Just when things were getting dire, I spotted two lean and fit-looking riders, a little older, with the sinewy, tanned arms and legs of experienced cyclists. They were wearing matching green and white jerseys, with a picture of a turtle and a slogan: "SLOW CYCLISTS MAKE FAST CYCLISTS LOOK GOOD—YOU'RE WELCOME."

As it happened, the Turtles were passing *me* at the time. But I liked their vibe, so I sped up slightly and tucked in behind them. I learned that their names were Butch and Max, and they lived and rode bikes in central Texas. Promising. And they had done the Hotter'N Hell more than ten times. Also promising. Butch was an ultrarunner, competing in races of 50 and even 100 miles, even in his sixties. Awesome. They knew the Hotter'N Hell, they knew a thing or two about pacing and endurance in the heat—and they were neither lagging nor hustling, just slowly and steadily getting it done. I vowed not to let them out of my sight.

The heat was like nothing I had ever experienced, let alone exercised in; it made my "heat training" rides in Utah seem like a dip in a cold plunge. The difference was the humidity. The air in Utah is so dry that it just gobbles up any and all moisture. Your sweat evaporates almost immediately, just as evolution intended; even on long rides, my skin would often be dry to the touch. Texas was a different story, sunny and steamy. Now I understood what the scientists meant when they described certain hot and humid conditions as "uncompensable"—meaning that most individuals would have difficulty maintaining a stable body temperature, particularly if they were exercising. Translated into English, it means, *this heat can kill you.*

And yet, weirdly, I felt okay. Far better than expected. I was surprised. I followed the Turtle Boys, mile after mile, as they calmly turned the pedals, ticking down the distance to the finish. They did not even appear to be sweating. The roads were almost but not quite flat, with gentle rolling hills; we would slow down on the shallow climbs, and speed up on the long, straight downhills. We began passing riders now, picking them off one by one—many of them people who had surged to the front in the gung-ho early morning hours. Now they were fading, their faces ghostly white with sunscreen and sweat and the cruel knowledge that fitness and heat tolerance are two different things. Soon, the pickups with their sag-wagon trailers began roaring past, full of some of these same riders.

The thought of quitting didn't even enter my mind. I was *going* to make it. I felt stronger with every mile. It was like an out-of-body experience. Who was this guy, riding calmly and steadily through the Texas Heat Dome? It was me. I was a Turtle Boy now, slow and steady, maybe not winning the race but abso-frickin-lutely finishing it.

At some point, we got the news that the ride was indeed being cut short due to the extreme heat. We had missed the cutoff time, so we were diverted back to Wichita Falls on an alternate route that was "only" 75 miles long. I felt more disappointed than relieved as we passed through the gates of Sheppard AFB, weaving around parked vintage warplanes and then passing down a broad avenue lined with cheering, high-fiving airmen. That picked up my spirits. I high-fived them back and attempted a Texas whoop of my own.

A Hotter'N Hell 75 would have to do.

We rolled into the back streets, crossed some railroad tracks, and then turned onto Scott Avenue one last time. And suddenly, there was the finish line in front of us, a big inflatable arch spanning the road. The announcer read our names, there were people clapping and cheering, and at long last I climbed off the bike and staggered over to stand with the Turtle Boys and dozens of other riders under a huge sprinkler, feeling the cool delicious water rain down on us, carrying away the dust and sweat and salt and pain from the hottest and most difficult thing that I had ever done.

Then I went looking for a Lone Star beer and some barbecue.

A week or so later, I returned to the Korey Stringer Institute for one more round of testing. It was the same basic deal: A long ride in the terribly hot heat chamber on the DUI-handlebar stationary bike, with Probey firmly inserted. I climbed on the bike and pedaled away—well, not actually "away," since it went nowhere. But going nowhere felt much better this time. After the Hotter'N Hell, the heat chamber felt almost refreshing.

My first session, in June, had all but destroyed me. What I remember most was the feeling of suffocating in the heat, struggling to contain my panic. It had required a maximal mental effort not to just quit right then and there. But now I spun the pedals with ease, smiling and joking with Huggy Bear and the lab assistants, Sean and David, as the minutes ticked by. I felt less thirsty, less tired, and much less hot than I had in June. It was like a ride in the park. (And if you still doubt the power of heat adaptation, just compare the before-and-after photos on this and the following page.)

Without a doubt, my overall fitness had improved after a summer of semidiligent bike riding. But not that much, it turned out: My time-trial pace was essentially the same as it had been in June. The main difference was that I could cope with the heat much better. Back in June, my heart rate and core temperature had begun climbing almost as soon as I entered the room, with no sign of a plateau. This time around, my heart rate was more than twenty beats slower, at the exact same power output and heat level that had crushed me in June. That represents a quantum leap in efficiency. It felt easy this time.

Heat testing at KSI, "before" heat training/adaptation—this is what not staying cool looks like. (Courtesy of the author)

I also felt cooler. According to Probey, my body temperature stayed about half a degree C lower as I rode than it had in June. In fact, it was lower from the very beginning of the ride, which seems like no big deal, but it actually was: A lower resting body temperature is one of the hallmarks of heat adaptation, as well as a positive marker of overall health—and possibly of longevity, too. (Weirdly, a lower resting body temperature may also indicate better mental health, as we will see in chapter 12.)

When I finished, we all traded high-fives. The heat training had worked! I was now officially heat-adapted, able to handle conditions

"After" heat training (and after surviving the Hotter'N Hell Hundred), with Sean Langan (c) and David Martin (r). (Courtesy of the author)

that, just a few weeks earlier, might have sent me to the hospital, or at least to bed for the rest of the day. And I had conquered the hottest Hotter'N Hell Hundred in years.

I had gone into the ride full of trepidation and fear. Right up to the day of the event, I was frightened of what might happen out there. But a relatively modest amount of heat exposure had worked magic. My body felt strong, but my mind felt even stronger—as Chris Minson had promised. I had not only endured the Hotter'N Hell, I had prevailed. For the first time in years, I felt like an *athlete*. All because I had spent a little bit of time doing something that I thought I didn't like.

But it wasn't just that I'd proved some kind of point. Or that I had lost my fear of the heat. It was bigger than that. As I would learn, in the coming months, heat adaptation can also save lives.

8

THE NICEST KID YOU'D EVER WANT TO MEET

> Sometimes the heat is my biggest opponent.
> —SERENA WILLIAMS

Right off the bus, Coach hit them with a "three-a-day," three tough workouts in the hot Florida sun, to kick off preseason training camp. It was hot when they hit the football field in the morning, and it was still hot when the last session ended after sundown. DJ Searcy was struggling. Close to midnight, an assistant coach found him passed out in the cabin where the kids bunked.

DJ, full name Don'terio Jacquavious Searcy, was going into his third season as a star defensive lineman for the Fitzgerald (Georgia) High School football team, the Purple Hurricanes. He was only sixteen, but he already weighed more than 260 pounds, huge for a high school player. His size struck fear in the hearts of opponents, though in truth he was anything but mean. "He was a gentle giant," says his father, Carlton Searcy. "He would lay kids out on the football field, and then he'd literally pick 'em up like they were a little child and stand them back up."

Colleges were already scouting DJ. His dream was to play for Clemson, and then maybe the NFL. Why not? The Purple Hurricanes—who everyone called the 'Canes—made States almost every year, and DJ was a star player.

But as the season drew near, DJ told his parents that he wasn't "feeling" football camp. He had spent the summer working for the Fitzgerald town recreation department, and he was out of shape. If you want to play, his father told him, then you must go to camp. That's the deal. "I told him, everybody looks up to you," Carlton Searcy says. "You can walk through the drills, but be present."

So he got on the bus with the rest of the team and rode down to Florida, to spend a few days training at a remote Bible camp near Gainesville. The morning after DJ got sick, he woke up early with his teammates for a rigorous series of calisthenics called The County Fair, then suited up for a midmorning scrimmage. He made it through the game before retreating to the cabin, saying he felt sick again. A little later, teammates found him passed out on the floor again.

A crowd of kids and coaches gathered around, unsure what to do. Someone called an ambulance, but the camp was located down a long dirt road, deep in the piney woods, so they loaded DJ into a pickup truck and sped to meet it. The EMTs found that he still had a faint pulse, and they raced for the nearest hospital.

Carlton Searcy had spent the night of August 1 in the field, bivouacked with new recruits. Born and raised in Fitzgerald, a quiet town of nine thousand surrounded by pine forests and pecan groves, the elder Searcy had played both football and basketball in high school, but he was best known for circulating a petition, as a student, to end the long-standing practice of holding segregated (and private) proms for black and white students. He joined the Army after graduation, not long after marrying Jaclyn, his high school sweetheart. The military was his ticket to a better life, and by 2011 he had risen to the rank of captain, in charge of a basic training unit at Fort Jackson, South Carolina.

It was so hot that week that Army guidelines limited the training that recruits could do during the daytime. So they took to training at night, rousing the troops at three or four o'clock in the morning for drills and maneuvers. It would still be 84 degrees out then, but cool enough to train. High school football players, however, had to practice whenever the coach told them to.

At a little before noon the next day, Searcy got a call from Jaclyn, from whom he was divorced but still cordial. She was screaming. DJ was in the hospital, she said; he better get down to Florida. Carlton jumped into his car, peeled out of camp, and sped south on I-95 at 100 miles an hour. An hour and a half later, his phone rang again. DJ is gone, Jaclyn told him.

He pulled over to the side of the road and cried.

"He was the nicest kid you would ever want to meet," Carlton Searcy says, by phone from Florida, where he lives now and works as a realtor. "And as I tell it, I live it over and over again. I was destroyed, man."

It took three months for the Ben Hill County school district to explain DJ Searcy's death. The superintendent attributed it to a pre-existing heart condition, citing "permanent heart damage that had resulted from years of uncontrolled high blood pressure," as though DJ were a sixty-year-old man. Head coach Robby Pruitt, a solid, good-old-boy type who remains a Georgia and Florida football coaching legend, added, "I don't know anything we could have done differently."

But in the end, DJ's death was part of a sequence of events that would lead to almost everything being done differently, at least in the world of Georgia football.

Incredibly, DJ Searcy was one of two Georgia high school football players to die from heatstroke on that very same day: August 2, 2011. Later that evening, a sixteen-year-old named Forrest Jones had also died, in a hospital in Atlanta. He had collapsed after the first day of preseason practice in late July and spent a week in the ICU before his liver and kidneys finally failed. "I just don't understand how it could happen to someone who is so strong," said his father, Glenn Jones, at the time. A third player, a sixteen-year-old named Isaiah Laurencin, had died in Florida the previous week, also from heatstroke.[1]

The three boys' deaths caused a public outcry, with calls for change. It was ten years almost to the day after Korey Stringer had died, at a

Minnesota Vikings preseason training camp in 2001. Since then, NFL teams, as well as the NCAA, had adopted stringent heat-safety measures to protect players.[2] But high schools, not so much.

Some states and school districts had heat-safety policies, but they tended to be vague, and no match for the hardcore, militaristic culture of Southern high school football, where coaches seemingly tried to outdo each other to see who could put their kids through the toughest practices—two-a-days, three-a-days, calisthenics until players puked, the whole nine yards.

"You gotta make 'em a man, so they're as tough as dirt," says Earl "Bud" Cooper, a former professor of kinesiology at the University of Georgia and an expert on heat safety for athletes.

But the deaths of DJ Searcy and Forrest Jones, and the events that followed, show how even a basic understanding of heat acclimation and heat illness treatment can be applied to save lives—not only of football players and other athletes, but of workers and potentially anyone else who must function, or even just survive, in hot conditions. And the changes that are required to accomplish this, the changes that make the difference between someone dying and them living, are almost absurdly simple and easy to implement.

Football is a traditional fall sport in the United States, but preseason practice typically begins in late July or early August, the hottest weeks of the year in the southeastern United States. Making matters worse is the fact that it is typically played by very large athletes wearing very heavy pads and helmets, going full tilt in the savage summer heat. Rob Huggins of KSI calls American football "a recipe for disaster, when it comes to heat-related illnesses."

But just because something is a "recipe for disaster" doesn't mean that people won't keep right on doing it. The occasional high school or college football player dying in summer practice was just something that happened every year, a brief story on the evening news, until a University of Georgia climatologist named Andrew Grundstein decided to take a closer look at the problem. Scouring news reports, death records, and any other source he and his colleagues could find, Grundstein identified and documented 61 high school football players who had died in the heat, nationwide, between 1980 and 2009. Since 1960, more than 120 had died.* And there had been many, many more nonfatal heat illnesses and heatstrokes—events that *could* have led to someone's death but luckily had not.³

One trend that immediately jumped out: During the 1980s and early 1990s, about one player per year had died, on average. But beginning in 1994, the pace had nearly tripled, to 2.8 deaths per year. It is tempting to blame the increase in deaths on climate change. But a separate group of researchers from the University of Alabama had reviewed weather data for the years 1984 through 1988 and concluded that even during that long-ago period, it had *never* been safe to play football in full uniform in the state of Alabama, during the first two weeks of August. Not one single day.⁴

Next, Grundstein and his team examined each individual death to determine its causes and conditions—the temperature and humidity, the time of day, the position or type of athlete, what they had been doing, and what they had been wearing, among other factors. Some patterns quickly became obvious: Linemen like DJ Searcy (and Ryan Swoboda) were most at risk. They were the biggest players, and they had been getting bigger, meaning they had more muscle mass to generate heat, and retained more heat thanks to their size. Linemen had comprised more than 95 percent of the total fatalities.

The players' protective equipment, their helmets and pads and long pants, contributed to the problem. While these XXL athletes were generating enormous amounts of metabolic heat in their bodies, and prodigious quantities of sweat to match, there was nowhere for that heat or sweat to go. Their equipment trapped much of the heat, while preventing their sweat from evaporating to cool them. Studies have found that football players' rectal temperatures will increase around 50 percent faster if they are wearing helmet and pads, as opposed to just shorts and a T-shirt.⁵ Thus, fully suited linemen can rapidly find themselves in danger at temperatures where they would be perfectly fine in lighter clothing.

---

* A 2024 study by Anderson et al found that a total of 159 youth, high school and college football players had died of exertional heatstroke nationwide between 1955 and 2021, but the authors had scoured news archives and similar sources. So the total number of reported deaths may not be accurate, because such incidents are not captured systematically in a single database.

The fact that football uniforms make players hotter had been shown in experiments dating back to the 1960s. In 2002, researchers from Penn State calculated that the safe temperature and humidity limits for football players wearing full uniform were considerably lower than those for athletes wearing only shorts and T-shirts. And nearly nine in ten player deaths had occurred at heat levels that exceeded those limits. But nobody had really paid attention until now.

The first few days of practice, typically late July and early August, seemed to be the most dangerous time by far. Nearly all deaths had occurred during the first two weeks of training, including DJ's and Korey Stringer's. And while football players had died in states from Florida to Minnesota, Georgia was the worst offender, with more player deaths than Alabama, Florida or Texas.

Clearly, Georgia needed to do *something* differently.

The miracle is that it did.

On a steamy Monday morning in late July, the metal locker-room doors clatter open, and the Marietta Blue Devils pile out—a whirlwind of teenage-boy energy in blue and white jerseys. Swinging their helmets and chewing their mouthguards, the hundred-odd players walk, jog, and swagger over to the practice field, where they fall in line for warmup drills. Soon whistles are blowing, pads are crunching, and footballs are flying. Some are caught. Some are not.

It's the first day of the second week of official high school football practice, all across the state of Georgia. The first Friday night game is a full month away, but things are already getting serious. "Fifty-one Mario! Fifty-one Mario!" yells head coach Rich Morgan,* his voice echoing off the tall trees behind the end zone. He is barrel-chested and very loud.

"NASCAR! **Duke**!" he screams. "Punch *RIIIGHT!*"

The players wheel into action, but the play doesn't go well, and the quarterback lobs a wobbly pass to an inside receiver who is immediately stopped for no gain. "We've got to *work* to get the quarterback's eyes! Work to get his eyes!" says Coach Morgan, before calling the next play: "Rodeo! *Rodeo!*"

They go again, marching down the field. Then they march back.

---

* Now head coach of the Niceville (FL) HS Eagles.

Whistles shriek. There is an awful lot of yelling. "You're telling me you didn't hear the play?" Coach screams at the QB, a few plays later. *"You didn't know the play!?!"*

His voice rises even more: "BECAUSE YOU USED THAT EXCUSE EARLIER!"

On the next play, a receiver wearing jersey #8 steps out of the action and staggers around at the side of the field, hands on his knees. His friend, who is injured and on crutches, spots him from the sideline. "Hey man, you gonna throw up?"

Number 8 shakes his head, unconvincingly.

The injured kid guides his buddy over to a training aide, who hands him a blue bottle of cold sports drink. He takes a knee for a minute or two and gulps from the bottle. In a blink, he's back in action, running to snag an off-target pass.

It looks like a typical high school football practice anywhere in America, but my guide, Marietta head athletic trainer Jeff Hopp, points out some subtle but important differences—all at least in part due to DJ Searcy's tragic death nearly fifteen years ago.

The most important change is the presence of a black Rubbermaid tub filled with water, sitting inconspicuously a few yards from one corner of the field. Research has found that if a heatstroke victim is cooled in ice water within ten minutes, their heatstroke becomes "one hundred percent survivable," in the words of KSI director Douglas Casa. Even Hippocrates, circa 400 BC, knew that quick cooling can save the lives of those stricken by the heat, although he did not have access to ice-making machines.

As soon as a player shows signs of potential heat illness, Hopp says, they will be stripped of their helmet and pads and uniform and dunked in the tub, while he and his assistants dump in buckets of ice from the ice machine in the trainers' office. They have not needed to do it yet, he adds, but that alone—just a tub and a few bags of ice—could have saved DJ's life, as well as that of Forrest Jones. "It is simple, but it takes time and a little bit of money, and a little bit of work," says Hopp. "That's really all it is."

The ice tub is one of several safety measures mandated by a complete overhaul of Georgia's high school athletic heat policy, which was written by a committee that included Bud Cooper and Jeff Hopp and implemented in 2012.[6]

Since 2008, the Georgia high school athletic association had been collecting data on heat illnesses in twenty-five school districts around the state, including Marietta. For three years, Jeff Hopp and his colleagues had kept track of every single heat-related illness at sports practices—not only deaths, which are exceedingly rare, but heatstrokes (also rare) and lesser incidents like heat-related syncope (fainting), dizziness, heat exhaustion, even heat cramps, all of which are more common. The researchers then correlated the incidents with weather conditions, sport and player position, body size, and the type of activity they were doing at the time.

They used that data to recommend specific safety guidelines that would become mandatory across every school district in the state. The ice tubs were only the beginning. The guidelines took aim at some cherished football traditions, such as the feared "two-a-days," twice-daily practice sessions, typically morning and afternoon. Meant to whip players into shape, two-a-days were also strongly associated with heat illnesses, because they exposed players to intense heat stress without giving them time to recover. The new guidelines struck a compromise: two-a-days were still permitted, but not on consecutive days. But three-a-days, the workouts that had led to DJ Searcy's death, were banned outright.

Other changes are less obvious but no less important. For example, while most of the players are wearing helmets and shoulder pads, Coach Morgan is working with a smaller group of players down at the 10-yard line—putting them through some highly unpleasant-seeming calisthenics, including crunches, up-downs, and sprints. These guys are wearing only T-shirts and shorts, because the guidelines now mandate that each player must complete five days of practice without helmets and pads before they put on full kit. The other players had put in their five padless days last week; these kids had missed a day or two.

This is for heat acclimation, just like with the South African miners. The point is to get the players used to working in the heat for a few days, *before* putting them in full uniform. The NCAA had mandated a similar five-day heat-acclimation period in 2003, and it had helped reduce heat illness rates. But it doesn't seem to have made practice easier.

By the second round of drills, the shirts-and-shorts players are looking pretty gassed.

"Might've been easier just to come to practice!" Coach yells when they're done.

One of the simplest, yet most important changes under the new guidelines is that coaches are now required to pay attention to the heat.

The Georgia data showed that most deaths had happened on days with high temperatures *and* high humidity. Mornings, with lower temperatures but higher humidity than afternoons, were surprisingly dangerous. Also, Andrew Grundstein's research had found that many of the player deaths from heat stress had occurred in weather conditions where, under safety standards used by the US military and numerous other sporting organizations, training would have been canceled outright. Yet, at the time, Georgia high school athletics had no such heat-safety guidelines in place.

Hopp flips open an app on his phone, which is connected to a sophisticated weather station mounted on a pole beside the field. Even that is novel: In the old days, practice was held in any and all weather conditions. Today is not overly hot, only in the mid-80s, but on hotter and more humid days, the new guidelines would limit the intensity and length of practices. And when it is too hot to practice safely, football practice (and every other kind of sports practice) must be canceled. Full stop.

In sum, it appeared that a few relatively simple data-driven changes could make a big difference in protecting the players. Selling the idea to hundreds of Southern football coaches was another matter. Going in to present the study to the Georgia High School Association board, which included five veteran football coaches, Bud Cooper expected that they would shoot it all down. "Our attitude was, we're going to tell 'em this stuff, and I don't know that they're gonna really like it," he says, "but it's the facts."

To his amazement, the coaches said yes to everything that was proposed—the deaths of the two boys, DJ Searcy and Forrest Jones, were still fresh on everyone's minds. The guidelines were put in place in time for the 2012 season.[7]

DJ Searcy at age fifteen, about a year before his tragic death. (Courtesy of Carlton Searcy)

Even more amazing: They worked. Over the next three seasons, Georgia high schools found that the incidence of heat illnesses dropped across the board, in some cases to zero. And there were no more deaths. In Georgia, at least, DJ Searcy and Forrest Jones were the last.*

Yet even as athletes and soldiers are protected from extreme heat, a much larger group of people remains at serious risk, especially in the United States: workers. These are the people who pick our tomatoes and deliver our packages, who roof our homes and repave our highways, who maintain (ironically) our solar-power fields, and so on. Seemingly, their safety is less of a societal priority than that of high school football players.

To understand what I'm talking about, Google "UPS driver passed out on Ring camera," and watch the video of a delivery man collapsing on a Scottsdale, Arizona, homeowner's front doorstep in 2022. The

---

* After a female high school basketball player in Atlanta died of heatstroke in 2019, when her coach made her run stadium sprints on a hot sunny day, the Georgia heat policy was expanded to include all sports, across the full school year, including competitions.

temperature that day was 110 degrees F. The company later said that the driver was "fine," in a statement. But at least two of this man's colleagues *have* died of heatstroke on the job in the last few years, and even in places like Arizona that suffer extreme heat every summer, most commercial delivery trucks are still not equipped with air-conditioning.

At the time, UPS said in a statement: "Our package delivery vehicles make frequent stops, making air-conditioning ineffective."[8] In 2023, UPS reached an agreement with the Teamsters union to have air-conditioning in all new trucks, starting in 2024. But as recently as May 29, 2025, the Teamsters were accusing UPS of "slow walking" the air-conditioned trucks, according to a union Facebook post.[9]

So far, we have primarily been talking about *voluntary* heat exposure—people choosing to do sauna or heated bathing, or heat training and heat adaptation for athletes. Now we are getting into the realm of *involuntary* heat exposure, which is a different and more serious situation. Ironically, we know much more about how athletes respond to extreme conditions than we do about workers—but it seems clear that some of the same heat-adaptation strategies used by athletes could be applied more broadly, to keep workers safe.

That it is dangerous to work hard in the North American heat has been obvious at least since the Great Depression summer of 1931, when fourteen workers perished of heatstroke while building the Hoover Dam in the desert near Las Vegas. Contemporary news accounts reported that temperatures at the dam site reached 120 degrees F (49 degrees C), in the shade; the 1930s, in fact, was one of the hottest decades on record in North America, with heat waves blanketing the United States year after year.[10]

Yet work on the dam continued at breakneck speed.

It took forty years for a nationwide, federal heat-safety standard even to be proposed, in 1972. Then, nothing happened for almost another forty years. A more serious effort to create workplace heat regulations (under the Occupational Safety and Health Administration or OSHA) finally began in 2011, under US President Barack Obama, but it faced serious industry opposition. The National Institute of Occupational Safety and Health published its own set of suggested (but not mandatory) heat-safety guidelines in 2016; meanwhile, OSHA was working on creating mandatory rules. But the first Trump administration put the rulemaking process on hold. Progress resumed under President Joe Biden, with KSI heavily involved in proposing and advocating new rules. Trump's second term brings more uncertainty, but as of this writing (July 2025), the federal government was still holding hearings and moving forward with the process.

Currently, only five US states—California, Oregon, Washington, Colorado, and (weirdly) Minnesota—impose even partial workplace heat-safety rules of their own, and those are limited in scope. Minnesota's regulations apply only to indoor settings, such as warehouses and factories, while Colorado's rules cover only agricultural workers. California has the most comprehensive set of heat-safety regulations, requiring employers to provide access to drinking water and shade or cooler spaces (for indoor workers), as well as mandating rest breaks and monitoring of employees during heat-acclimation periods. But even those are somewhat vague, simply directing employers that they must *have* "high-heat procedures" in place, for workers in temperatures above 95 degrees F, without specifying what those procedures should be. So while Georgia protects its 30,000-odd high school football players with well-defined standards and procedures, California gives no guarantees to its nearly six million workers, many of them in agriculture, who are potentially at risk from high heat.

Another issue is that we don't really know how dangerous it is to work in the heat, because the data about heat-related injuries and deaths among US workers are so vague and unreliable. Between 2017 and 2022, some 1,054 heat-related illnesses were reported to OSHA, including 211 fatalities. But KSI experts say those numbers likely represent just a fraction of the actual toll—and likely omit deaths due to heart attacks or other causes that are related to excessive heat.

"It's the Wild West," says Rob Huggins. "There is massive, massive underreporting of deaths due to heatstroke. They might call them 'cardiac events,' they might call them 'liver failure,' they might call them 'kidney failure'—when in actuality it was the exertion in the heat on the farm, or on this megastructure they were working on, or being out in solar fields where it's hot beyond hot, and people are subjecting themselves to that."

So while middle- and upper-middle-class people stroll into their hot yoga classes carrying $100 crystal-infused water bottles, then hop over to Whole Foods to pick up some organic produce, farmworkers in South Texas are unable to even take water breaks while picking that produce in the hot sun. This has contributed to a mysterious rise, among agricultural workers in the southern US and Central America, of something called "chronic kidney disease of unknown origin"—a condition that some studies trace to chronic heat stress and dehydration among people working outside in the heat.[11] Hotspots of chronic kidney disease have been identified in agricultural areas of California and the Mississippi River valley, as well as in Mexico and Nicaragua.

Rather than protecting these essential workers, some places such as Florida and Texas are even rolling back what limited worker protections they do have. The Republican governor of Florida, Ron DeSantis, signed a bill in 2024 banning local municipalities from imposing their own heat protection measures for workers. Texas Republican Governor Greg Abbott had approved a similar law in 2023, rescinding local regulations in Austin and Dallas that had merely mandated water breaks for workers—one 10-minute break every four hours. Those were the only two cities in the entire state that had imposed even minimal safety requirements, but the new Texas law also forbids any other localities from adopting similar measures.[12]

The US lags far behind other nations in terms of worker protections in the heat. The European Union has somewhat vague suggested guidelines, urging water breaks and limiting work on the hottest days, but China is more specific, limiting outdoor work at temperatures above 99 degrees F (37 degrees C), and requiring work to be suspended at 104 degrees F (40 degrees C) and higher.

Even totalitarian nations in the Middle East do more to protect workers than Texas does. In Qatar, thousands of foreign workers reportedly died during the construction of stadiums for the 2022 FIFA World Cup, according to human rights groups. Workers were forced to labor in 105-degree heat, and many were housed in un-air-conditioned trailers and even shipping containers. But Qatar now bans outdoor work from 10 a.m. to 3:30 p.m. between June and mid-September, the very hottest months. Other Persian Gulf nations, such as Kuwait, the United

Arab Emirates and Bahrain, also ban or limit outdoor work in some way during the summer months.

But workers in the United States remain in a precarious situation with zero safeguards. The standard argument is that worker-protection measures impede productivity and ensnare businesses in needless red tape. As the Republican state legislator who introduced the Texas bill explained in a press release, "progressive municipal officials and agencies have made Texas small businesses jump through contradictory and confusing hoops."[13]

Another argument is that providing safety measures like shade and cooling tubs is too costly and impractical, particularly in agricultural, utility, and delivery applications. This is a problem that technology may help solve: A $500 device called the ColdVest, which looks like a life vest but is filled with chemicals that become ice cold on contact with water, could help cool down heatstroke victims in places where ice is not available.

There is at least some evidence that heat-safety measures may even *improve* productivity, as the South African gold mine owners learned many decades ago. At a sugarcane mill in Nicaragua, for example, implementing a voluntary system of rest and hydration breaks for workers either increased or did not harm their overall productivity, according to one study—while significantly reducing the incidence of heat injuries and kidney disease among the workers.[14]

"One of the three leading causes of death of laborers is heatstroke," says Doug Casa of KSI. "I think that the number right now is fourteen million Americans work in the heat each day. And we have no heat standard. There are standards right now for burn safety and head injuries and helmets and falls, but there's no heat standard in America right now for workers. They don't even have to provide fluid-cooling strategies. It's totally nuts."

The lack of attention to worker safety in the heat reflects, I think, a broader lack of understanding of heat and its dangers—and our own lack of exposure to the outdoors. Because it is chilly in the produce aisle of the grocery store, and nice and cool in our suburban homes and city apartments, we don't even think about how hot it might have been for the people who picked the strawberries, the watermelons, or the tomatoes.

But as the story of Georgia football shows, the science of heat acclimation works. It has saved the lives of athletes and it could help protect workers in hot climates around the world if properly applied. And as we will see in the next chapter, heat adaptation can even help city dwellers become safer as our world gets hotter.

9

TOO DARN HOT

> I'd like to sup with my baby tonight
> Refill the cup with my baby tonight
> But I ain't up to my baby tonight
> Cause it's too darn hot
> —COLE PORTER

WHY do we care about American high school football players? Because they are vulnerable children at risk of serious illness, injury, and even death, for starters. But they are also explorers, in a sense—climate explorers, already operating in the hotter world that some scientists believe awaits the rest of us, decades from now. American football is therefore like a natural experiment, to see how far the human body can be pushed in extreme heat.

A Penn State study suggests that the effective temperature experienced by an American football player wearing full pads and helmet is several degrees warmer than for someone wearing lighter clothing.[1] Understanding how to protect football players can therefore help the rest of us understand and anticipate the most severe possible effects of climate change, not only on athletes but on workers, the elderly, and anyone else who must survive in extreme conditions—including, possibly, most of the rest of us.

Earlier, we saw how relatively minor changes in procedures—modest amounts of heat adaptation, coupled with greater attention to environmental conditions and hydration needs, and some basic safety precautions—can make even American football players much safer in the heat. Workers too, if and when we muster the political will to do the right thing.

But what about the rest of us?

In a 2010 paper published in *Proceedings of the National Academy of Sciences (PNAS)*, two atmospheric scientists named Steven Sherwood and Robert Huber declared that they had calculated the absolute upper limit to human survival in the heat: 95 degrees Fahrenheit, or 35 degrees C.[2]

They were not talking about temperature alone, but a measure called "wet-bulb temperature"—literally, the temperature reading you would get if you covered the bulb of a thermometer with a wetted cloth and exposed it to a stiff breeze. Strange as this technique sounds, the idea is that this wet-bulb temperature reading, or WBT, accounts for the evaporative cooling power of sweat, and thus more accurately reflects the heat stress that a person might experience at a given temperature and relative humidity.

The WBT is somewhat similar to the heat index, which most people are familiar with. The heat index (and variations such as "RealFeel") combines temperature and humidity to estimate how hot a given set of conditions might feel to a real person. Specifically, it is meant to reflect the heat stress experienced by a 145-pound, five-foot-six-inch person, wearing short sleeves and long pants, walking at 3 miles per hour in the shade.* Thus a 95-degree Fahrenheit day in Washington, DC, with 50 percent humidity, yields a heat index of 105 degrees F—meaning it feels hotter than the thermometer reading. But the same temperature in Salt Lake City, at 15 percent humidity, yields a heat index of only 90 degrees F, which helps explain why I left DC.

While weather forecasters like the heat index because it is so easy to use and understand, scientists prefer WBT and its outdoor equivalent, Wet-Bulb *Globe* Temperature (which considers both wind and the heating effect of the sun), believing that those measures more objectively

---

* Simple as the heat index seems, the equations used to define it, in the original 1978 paper where it was proposed, run for more than ten pages.

describe the thermal environment.* But because the wet-bulb temperature is *always* lower than the actual, ambient temperature (also known as the "dry-bulb" temperature), WBT is much trickier to communicate to the lay public. A temperature of 95 degrees F (35 degrees C) does not sound all that hot to those of us in North America, Australia, the Indian subcontinent, and many other warm places around the world where 95 degrees F represents a relatively common summertime high temperature. But in fact a *wet-bulb* temperature of 95 degrees F (35 degrees C) represents a truly perilous level of heat and humidity. (Imagine 110 degrees F at 55 percent relative humidity, for example.)

According to the authors of the *PNAS* paper, in such conditions even healthy young people would no longer be able to cool their bodies by sweating. As if that were not sufficiently dire, they also warned that as climate change progresses, "the area of land likely rendered uninhabitable by heat stress would dwarf that affected by rising sea level. Heat stress thus deserves more attention as a climate-change impact."[3]

The authors made clear that this represented the most extreme possible scenario of climate change, produced by burning all of the Earth's fossil fuels over the next three hundred years, and that such conditions had rarely been observed (yet). Nevertheless, their alarming paper made headlines. One widely-read article in *New York* magazine, later expanded into a bestselling book called *The Uninhabitable Earth*, emphasized the authors' dreadful prediction that up to 20 percent of the world's population could be facing death from excessive heat—or as the article vividly put it, being "cooked to death from inside and out."[4]

But when physiologists took a closer look, they found that it did not withstand experimental scrutiny—and that it masked a far more nuanced story about heat, humans, and thermal adaptation.

The theoretical 95 degrees F WBT survival limit turned out to have two serious flaws as a model. For one, it did not appear to be remotely accurate, as applied to actual humans. When Penn State scientists subjected volunteers to equivalent levels of heat and humidity in their heat

---

* WBT is typically only used to describe air temperature and humidity in *indoor* settings, such as homes, workplaces, and heat labs; in outdoor environments, scientists use the "Wet-Bulb *Globe* Temperature," or WBGT, a modification of WBT (devised by the US military in the 1950s) that takes into account the potent effect of the sun.

lab, they found a wide range of responses, rather than one single limit for everyone. While they were (obviously) not allowed to push their study subjects to the point of heat illness or death, they found that their young, healthy subjects ran into difficulty well below the theoretical 95 degree F "limit"—even as low as 81 degrees F WBT, which is about what it was on the August day when I observed football practice in Georgia.[5]

It depended on the person's individual physiology, their level of physical activity and their physical fitness. Age was another important factor:[6] Older subjects struggled in the heat, which is why the elderly are far more likely than the young to perish in heat waves. But nearly everyone seemed to have more problems in high-humidity situations than in dry heat, says Kathleen Fisher, a Penn State PhD student who participated in the research. In other words, Florida hot took more of a toll than Arizona hot, despite far higher temperature readings in the latter.

But the results were so divergent that it cast doubt on the entire idea that some arbitrary heat limit even exists. "It's more of a spectrum," Fisher says.

Another, more significant issue was that people were *already* dying in heat waves around the world—in far milder conditions than the supposed 95-degree-F WBT threshold. Heat waves in Europe in 2023 killed tens of thousands of people, in temperatures in the 90s or low 100s (F) that would scarcely faze an Arizonan, a pilgrim to Mecca, or a rider in the Hotter'N Hell Hundred. A weeklong stretch of severely hot weather in Chicago in July 1995, when the heat index hit an unbearable 126 degrees Fahrenheit, had killed hundreds of people. Yet far hotter heat waves in India and Pakistan had not caused nearly as many deaths, it seemed.*

Why was that?

Adaptation.

It turns out that some of the same ideas we have been exploring—the variability of heat tolerance, the role of heat exposure and relative fitness,

---

* To be fair, the *New York* writer also wondered about this, in a 2022 op-ed in *The New York Times*. https://www.nytimes.com/2022/07/14/opinion/environment/heat-waves-india-pakistan-climate-change.html.

and basic strategies of heat adaptation (and even heat shock proteins)—also become relevant in protecting humans from a shifting climate. Not that we can adapt our way out of climate change, but our unique physiology can play a role in, possibly, helping it suck a little bit less.

Because we humans are highly adaptable creatures, when it comes to climate as well as in other domains of life.

"I think if your perception was that we're going to warm the planet, and then there will come a point where we just can't do it anymore and we keel over dead, en masse, in some part of the world—that's not accurate," says David Romps, a professor of climate physics at the University of California at Berkeley who has also questioned the theoretical 95-degree-F (35-degree-C) threshold.*

"On the other hand, even before we get to these hyperthermic and fatal [temperature] values, a lot is going to change about how we live, because we will make adaptations. We will stay indoors, in air-conditioning. We won't go outside and work on hot roofs in the middle of the afternoon during a heat wave. There will just be things that we don't do. So I think about it less as *we're all going to die* and more as *we are going to be forced to adapt so that we don't die*."

Which humans and human-like species, notably the Neanderthals, have already been doing, for tens or even hundreds of thousands of years. Thousands of years ago, for example, the southern Sahara Desert (in what is now part of Niger) was a lush, green and agriculturally productive place, with thriving communities and lakes full of fish. When the climate shifted, and the area became hotter and drier, people simply moved away, leaving behind tools, pottery, and burial sites that puzzled archaeologists before they figured out what had likely happened.[7]

In other hot regions, people simply adapted their lifestyle to the warming climate. Desert-dwellers in North Africa, Southern Europe,

---

* Romps and his colleague, Yi-Chuan Lu, both in the UC Berkeley Department of Physics, also questioned the validity of using WBT and WBGT for physiological purposes. They suggest that heat index is actually a better predictor of how humans will handle a given level of heat and humidity—because unlike WBT or WBGT, the heat index is at least based on human physiology. ("Wet bulb temperature describes the temperature of a dead, sweaty person in a stiff breeze," Romps scoffs.) They applied heat index to Huber and Seymour's climate scenarios, and found a far less dire result, in which only a small fraction of the world's population would truly be at risk of death from excessive heat. (See e.g., Lu and Romps, "Extending the Heat Index," JAMC 2022.)

and the Middle East developed light, loose styles of clothing that allowed for air circulation (and thus cooling) around the body; they built cool, open dwellings with shaded courtyards and thick walls; and they wisely stayed out of the sun in the middle of the day, a custom that (barely) survives as the *siesta*.

We evolved in the heat, in tropical/equatorial climates that would likely seem unbearable to modern humans. So it seemed a fair trade to surrender our fur in exchange for the superpower of sweating; cooling was that important. As a result, our happy place climate-wise is to be naked in about 82-degree-F (28-degree-C) weather—such as the Rift Valley of East Africa, or, say, Hawaii.

We are less well suited to cooler climates than to warmer ones, so as our ancestors began to migrate to higher latitudes, they devised clever adaptive strategies to survive—living in caves, creating dwellings from the materials at hand, including snow and ice, wearing the skins of other animals, and harnessing fire (possibly to heat their primitive saunas).

Their physiology changed subtly too. People whose ancestors had migrated to colder climates tended to develop larger bodies, over many generations, with more muscle mass to generate and retain heat, and more fat storage for sustenance and insulation. Their hair became thin and fine, as it was no longer as important for protecting them from the heat of the sun. Even the shape of their faces changed, with longer and narrower noses allowing for humidification and warming of the colder air they breathed.

Some of these adaptations reached all the way down to the subcellular level: The biologist Doug Wallace from the University of Pennsylvania, a world expert on mitochondria (the tiny little machines in our cells that help create power, among other things), has found that Arctic peoples such as the Inuit may generate more metabolic heat within their cells than people of African descent,[8] because their mitochondria have evolved to be less efficient (and thus generate more "waste" heat, which serves the secondary purpose of keeping their owners warm).*

---

* Mitochondria have their own small genome, distinct from ours, reflecting their origin as separate organisms that migrated into single-celled creatures about a billion-ish years ago;

Along the way, they developed formidable tolerance to cold as well as heat. When Charles Darwin landed at Tierra del Fuego, the chilly southern tip of South America, he and his HMS *Beagle* shipmates were astonished to encounter Indigenous people who were almost entirely naked; the clan elder who greeted them wore only the skin of a guanaco (a furry four-legged animal like a llama), draped over his shoulders like a cape. And nothing else.

This thermal flexibility enabled humanity to spread across the globe, from the Arctic to the tip of South America—by adapting to colder temperatures as well as warmer ones. We even survived a one-hundred-thousand-year-long string of ice ages that ended a mere eleven thousand years ago. Our worst enemy was the cold—and it still is. Even now, the greatest climatic threat to human life is not heat, but cold. Epidemiologists have compiled mortality data from across the globe and discovered that far more people appear to be dying due to colder temperatures than warm conditions—even in places that are affected by extreme heat. This is pretty much the opposite of what you typically hear on the news, but it's true.

While relatively few people die as a *direct* result of either heat or cold, from heatstroke or hypothermia, there is a significant and inescapable correlation between temperature and what the epidemiologists call "excess deaths"—that is, more people dying than expected. Up to 10 percent of all deaths are thought to be related in some way to what scientists call "non-optimal temperatures." While heat or cold was not a direct cause, it may have been a contributing factor. And for every death associated with excessive heat, about nine people die from causes linked to cold. The ratio of cold-related excess deaths to heat-related deaths is nine to one. It's not even close. And while this balance is slowly shifting due to climate change, cold remains the greater threat to human life.[9]

But it is the geographical distribution of heat- and cold-related mortality that is most telling. According to a massive analysis of global

---

this mitochondrial DNA is less stable than our own genome, and thus is more prone to mutation–and more rapid evolutionary change. In other words, some scientists think mitochondrial evolution progresses more rapidly than organism-level evolution.

mortality and temperature data, compiled by dozens of epidemiologists and climatologists and published in the *Lancet* in 2021, the region with the highest rate of excess deaths linked to cold temperatures is sub-Saharan Africa. There are more cold-related excess deaths per capita in Nigeria than in Boston.[10]

Conversely, we find greater heat-related mortality in Europe than in stereotypically hot places such as Africa or Indonesia. Indeed, people in Europe have been dying from (or in conjunction with) heat in conditions that would barely cause an Indonesian to sweat. And sub-Saharan Africans do just fine in weather so warm that it would drop Bostonians in droves. Around the world, the pattern is the same: People in hot regions are far less sensitive to the heat than those from cooler climates, for the very simple reason that they are used to it. This is the real reason why heat waves in Europe or the Pacific Northwest of the US are so dangerous: The locals simply aren't accustomed to the extreme, sudden heat.

This is also why it can be dangerous to travel from a cool climate to a very warm climate without allowing oneself a period of acclimation. In March 2014, for example, my friend the writer Matthew Power flew from New York City to Uganda, where he planned to join the British explorer Levison Wood on part of his 4,000-mile trek down the length of the Nile River basin. The temperature in New York was in the 30s, while Uganda was well over 100 degrees F. Matt wasted no time and got on the trail the day after he landed. He was a seasoned journalist and an experienced adventurer, who had journeyed from his Brooklyn townhouse to such hot and steamy locales as the Amazon jungle and Thailand, after the 2004 Asian tsunami. He was not soft. But on the third day of walking on rough trails in the heat, he began feeling faint and overheated. Soon, he struggled to keep water down. After noon, he passed out. There was no way to cool him, and the team was not able to arrange a rescue helicopter in time. He died of heatstroke, leaving behind a heartbroken fiancée and many sad friends and colleagues.[11]

Matt's story reminds us of the sneaky danger of heatstroke, and how it can so easily trap the unprepared. But the key factor was not his hydration, clothing, or physical fitness, but his level of *acclimation*: He had gone from frosty Brooklyn to steamy Uganda in the space of 24 hours, without giving himself even a day to acclimate. Anyone traveling from sea level to, say, Everest Base Camp (at 17,600 feet) would be urged to do so gradually because of the change in altitude, but heat acclimation is still a novel and largely unfamiliar concept, to most.

Matt Power's sad story illustrates a larger point: Many of the dangers attributed to climate disruption, including cases of deaths of hikers in places like the Grand Canyon and Canyonlands National Park, are perhaps better explained in terms of heat acclimation (or the lack thereof).

This turns out to be true at both the individual level and more broadly. In another paper, some of the same scientists cited just above drilled down to the level of individual cities, across a broad range of countries, from Thailand to Australia to Canada to Italy, comparing mortality statistics and daily temperature data for each locale. For each individual city, they plotted excess mortality versus daily temperatures—and determined the "optimal temperature" for each, a level of heat (or cold) where excess mortality was the lowest. To their surprise, this optimal temperature varied enormously from city to city.[12]

For example, the excess-mortality rate in Bangkok bottoms out at about 86 degrees F (30 degrees C), as you can see by the dotted vertical line in the graph on the following page. But that same temperature would cause deadly havoc—more than doubling excess deaths—in London (middle graph, next page), where the optimal temperature is just 67 degrees F (~19 degrees C).

Some of these variations can be explained by differences in infrastructure, lifestyle and living habits, as well as the availability of air-conditioning (or heating). Bangkok was obviously built for tropical conditions, and air-conditioning is quite common there (roughly 80 percent of homes). In London, homes were built to keep heat *in*, guarding against the cold and damp English weather, while only about 5 percent have air-conditioning—making it uniquely vulnerable to extreme heat events that can strike quite rapidly and unexpectedly.

Air-conditioning plays a complex role. On the one hand, it can make the difference between life and death for certain vulnerable groups, as heat waves in the Pacific Northwest and in Europe have shown. But

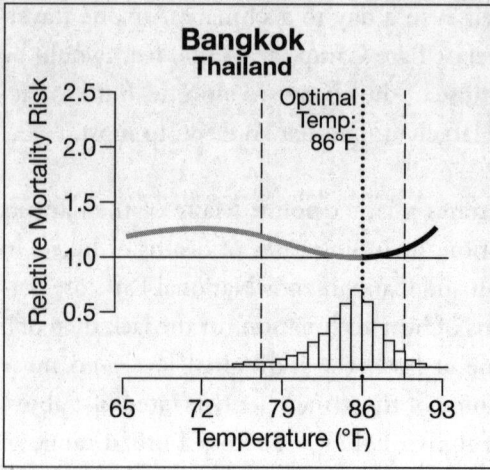

Temperature in Celsius vs. mortality risk for Bangkok. Note that the "optimal temperature" is indicated by the vertical dotted line, while the gray and black curve represents relative risk of mortality, or RR, at temperatures cooler and warmer (respectively) than the optimal temperature.

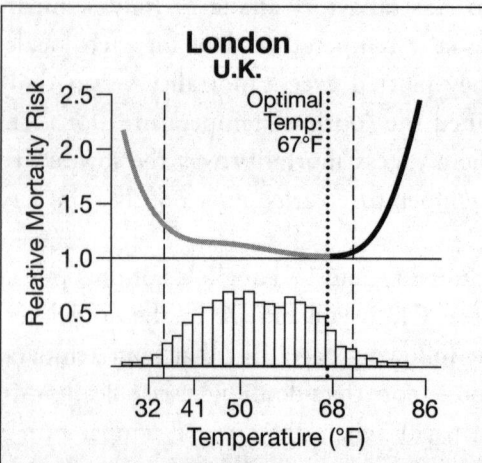

In London, the optimal temperature (lowest mortality) is about 67 degrees Fahrenheit, nearly 20 degrees cooler than Bangkok—but notice how the black curve rises steeply to the right. This means that extreme heat—even temperatures in the 80s Fahrenheit—is more acutely dangerous to Londoners than residents of Bangkok; fortunately, such hot days are relatively infrequent.

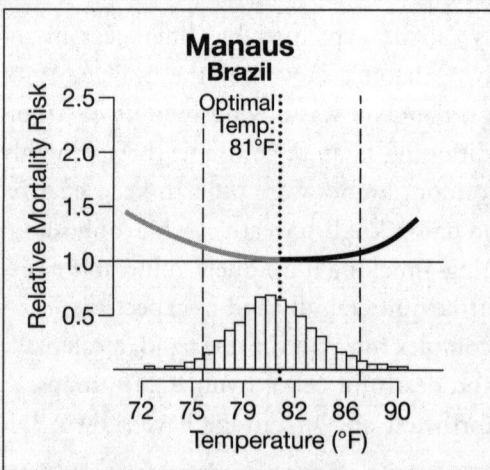

Manaus, in the Brazilian Amazon, is hot and humid like Bangkok, but with lower prevalence of air-conditioning. Even so, the optimal temperature is around 81 degrees Fahrenheit, with less heat-wave-related mortality than London. (Graphics from Gasparrini et al, 2015, *The Lancet*)

air-conditioning can also make some people *more* vulnerable to heat stress, as one fascinating Chinese study found. Researchers in Beijing compared people who worked in heavily air-conditioned buildings with those who worked in un-air-conditioned ("naturally ventilated") office buildings and found that the non-AC folks fared far better on a heat-stress test, similar to the one that I did at KSI.* The authors wrote, "It appears that long-term exposure to stable AC environments may weaken people's thermal adaptability."[13]

But air-conditioning doesn't explain all of the difference in temperature-related mortality. Deep in the Amazon rainforest, the city of Manaus, Brazil, is extremely hot and humid—and less air-conditioned than Bangkok (only around 20 percent of Brazilian homes have air-conditioning). Yet the optimal temperature for Manaus, mortality-wise, is still a relatively warm 81.5 degrees F (27.5 degrees C).

"As far as the epidemiologists are concerned, there's a real mystery as to why we seem to be so acclimated to various temperatures, because it's not just explained by air-conditioning prevalence and things like that," says Patrick Brown, a climatologist and senior fellow at the Breakthrough Institute, a climate think tank. "So it seems like there actually is some physiological adaptation."

In other words, Bangkokers and residents of Manaus are naturally more heat adapted than Londoners, which makes sense: It's generally hotter in those places, on any given day.

The role of natural heat adaptation in health outcomes has been borne out in other research. A large Japanese study looked at nationwide hospital data and found that patients who were admitted to hospitals with heat illnesses in generally cooler parts of the country (those with a lower average WBGT or wet-bulb globe temperature), tended to be 32 percent more likely to die in the hospital than those in warmer regions.[14] People who lived in warmer climates were more resistant to sudden heat stress.

Think of it like home-field advantage in sports. The Minnesota Vikings are an American football team from the northern Midwest, and because their climate is so cold in fall and winter, they play their home games in an indoor, climate-controlled stadium. (Their practice facility

---

* Also, while this was a small and non-statistical study, the non-AC workers were also generally thinner, the authors noted.

has both indoor and outdoor fields.) But when they travel to Florida to play against the Miami Dolphins, about 1,500 miles (and nearly 20 degrees of latitude) to the south, they face a distinct disadvantage in the Dolphins' steamy, outdoor Hard Rock Stadium. The Dolphins players, accustomed to playing and practicing in subtropical heat and humidity, feel fine.* Their built-in heat adaptation gives them a slight but definite edge. To combat this, some northern teams (such as the Philadelphia Eagles) have begun prescribing sauna sessions for the players during the days leading up to away games in Florida.

The stakes are greater than the outcome of a mere football game. People who live in Miami are acclimated to Miami's heat and humidity. Residents of Minnesota, on the other hand, are not—and when Miami-like temperatures strike the Upper Midwest, as they did in July 2001, it puts many people at risk, even athletes.

It was so hot in Minnesota, on July 31, 2001, that farmers were urged to keep their livestock indoors, in the shade. But the Minnesota Vikings still held practice outside, as planned. On July 31, five Vikings players collapsed during practice in the heat, and one of them was Korey Stringer, then a twenty-seven-year-old offensive lineman who sweated so much he carried a towel around with him wherever he went. A novice athletic trainer took him to an air-conditioned trailer to try to cool him down— not knowing that the only way to stop a heatstroke is by immersing the person bodily in a tub of cold water, within half an hour. Air-conditioning won't do it. (Nor will cold towels or icing hands, feet, and armpits.)

By the time the ambulance arrived, his core temperature was nearly 109 degrees F, and there was no coming back. He died early the next morning of multiple organ failure—ten years almost to the day before DJ Searcy.**

---

* Making matters worse is the fact that, as the stadium was originally designed, the Vikings (and all visiting teams) were forced to sit in the full Florida sun during afternoon games, while the Dolphins luxuriated in the shade. The Dolphins have since added a partial roof and have also agreed that all games during the still-hot month of September must start at 4 p.m. or later, to avoid playing in the full heat of the sun.
** Later it turned out that Stringer was trying to lose weight by, among other things, taking the supplement ephedra—which is now banned due to numerous deaths associated with its use. Stringer's widow, Kelci, would go on to help found the Korey Stringer Institute at UCONN.

As our climate becomes more uncertain and unpredictable, perhaps some of these same heat-adaptation strategies we have been talking about for athletes and workers, perhaps even judicious use of saunas, could help make us all more resilient, more able to tolerate unexpected heat stress.

There is already some research suggesting that modest and easy heat-acclimation methods could help many people, especially older adults, who typically account for the majority of excess deaths in severe heat waves and from heat illnesses in general (as well as excess deaths from cold). Something about the aging process seems to weaken our thermal adaptability, narrowing our range of tolerable temperatures. Individuals with diseases such as Parkinson's, as well as depression, seem to do even worse—sweating less, and feeling hotter. Their thermoregulatory systems simply don't work as well, and as a result they tend to run into trouble—and greater risk of heat illness—at lower temperatures than younger people, says Kathleen Fisher.[15]

Can confirm: One reason why I struggled, in my first session in the KSI heat lab, is because in my fifties, I am less thermally nimble than I may have been in my twenties. Even Bode the Dog, now past twelve years old, enjoys hot days much less than he did just a couple of years ago. But a deeper dig into the data revealed that age is not the only risk factor at play—and that the danger can be mitigated with the proper tactics.

The primary danger from heat, to young and old alike, it turns out, is actually not the likelihood of *direct* heat illnesses like heatstroke, says Anthony Bain, of the University of Windsor. Rather, heat stress creates its own, downstream physical stresses, beginning with cardiovascular stress. "Rates of cardiovascular events increase dramatically during a heat wave," he says, "while there are very few deaths that are actually associated with a heat injury [e.g., heatstroke or heat exhaustion]."

This is also why heat-related deaths, for example of workers, are so difficult to track. They do not always present as heat-related. "They're all labeled as a cardiovascular event, which they are," says Bain, "but it's really the heat, because it's such a big cardiovascular strain."

Yet one group of older adults appears to be more or less immune from heat stress: masters athletes, who rank among the fittest of the fit for their age. In multiple studies over three decades, Penn State researchers and others have found that highly aerobically fit older men and women, in the higher percentiles of VO_2 max for their age, fare far better in the heat than sedentary, less fit people of the exact same age. In head-to-head comparison studies, very fit older athletes have proved to be every bit as heat tolerant as people decades younger than them. They even sweated more, like younger people do. The higher their VO_2 max, regardless of age, the better they felt in hot conditions.[16]

Heat training can potentiate this effect. Another small study from 2014 found that a group of fit cyclists in their fifties and sixties responded nearly as well to a weeklong heat-acclimation protocol as a group of similarly fit athletes in their twenties and early thirties. "These data suggest *age does not affect the capacity to acclimatize to exercise-heat stress in highly trained adults* undergoing short-term heat acclimation," [emphasis added] the authors concluded. In other words, high fitness plus heat training is like a fountain of youth.[17]

While no amount of heat acclimation can make climate change go away, as we prepare for a likely warmer future (or even just next summer), nearly everyone—not only athletes—can benefit from basic strategies of heat adaptation. Even small improvements in aerobic fitness, plus a bit of heat acclimation, can help make those heat waves, now a fact of life in many places, a little less miserable—and less deadly.

These relatively simple interventions are easy to implement and access. For example, in Japan, a group of sedentary sexagenarians exercised on stationary bikes four times a week for four and a half months, improving their VO_2 max by about 20 percent, which is a significant but not earth-shaking improvement in fitness. By the end of the training period, they were also sweating more, and they began to sweat at a lower body temperature. They were *sweating better*, because they had become more heat-adapted as well as more fit. Other studies have found similar benefits for older people who complete short-term heat-acclimation protocols. (Although older men, in particular, tend to be less thirsty and thus more prone to dehydration.)[18]

While aerobic fitness seems to be a prerequisite, specific training for heat acclimation is the critical piece, says Bain. "How do we protect those most vulnerable to heat-related injury? Aside from avoiding the heat if you can, the best way to protect yourself is to acclimate, so your thermal tolerance is improved."

We know that one of the best ways to prepare for exercising in the heat is by . . . exercising in the heat. But not everyone can exercise intensely, especially in hot conditions; more to the point, we want to become heat-adapted *before* the heat wave strikes. This is where sauna and other forms of heated bathing can play a role. Research has found that even light to moderate exercise at room temperature, followed by an equivalent time in a hot tub or hot bath, can induce potent heat adaptations. A UK study from 2021 found that 40 minutes of easy jogging at 66 degrees F (19 degrees C), followed by 40 minutes of immersion in hot water at 104 degrees F (40 degrees C) induced greater heat-adaptation changes after six days than an hour-long daily treadmill run at a sweaty 91 degrees F (33 degrees C).[19]

Make no mistake: Forty minutes in a 104-degree-F hot tub is a long, hot session. But jogging in 91-degree-F heat is possibly even worse. Pick your poison. Other research has suggested that the jogging part might even be optional.[20] Multiple studies have found that sitting in a warm bath for an extended period can induce some heat-adaptation changes, even in older adults; one small pilot study found that simply soaking older women's feet and lower legs in a hot bath helped improve their heat tolerance.

Anything that warms up the body can help with heat acclimation, Bain says. But you can't half-ass it; you need to feel at least some degree of heat stress. "You *have* to get your core temperature up, by a degree C or so, and keep it there for an hour," he says. "You need a certain dose to have a measurable effect, and that's where I think many people fail, where they just do a little bit."

The upside: Getting a little uncomfortable now can prevent a great deal of discomfort, or worse, later. And one of the most effective tools for heat acclimation is also one of the most ancient and most popular, not to mention fun: Finnish sauna.

III
HEALING

10

STEAMED

> The thoughts and feelings that emerge from being on the sauna bench could never appear anywhere else in the world.
> —F.E. SILLANPÄÄ, Finnish author and Nobel laureate

As summer faded into fall, the excitement of the Hotter'N Hell gave way to melancholy. Normally, I celebrate the onset of sweater weather; my middle name should be "Cashmere." But Texas had changed me. For the first time, I actually missed the summer heat.

All of a sudden, it seemed like everyone was talking about sauna. According to every fitness and health guru on the internet, sauna was *extremely healthy*, and the hotter the better. My Instagram feed filled up with clips of influencers—always ripped, often tattooed, typically scantily clad—describing their sauna and cold plunge routines in exacting detail: X number of minutes in a sauna at a temperature of Y degrees, followed by a dunk in a tub filled with ice water, capped off by a dripping-wet fist-bump. And, of course, a lengthy discussion of the health benefits they had thereby achieved.

These were all variants of a species known as the "Sauna Bro," a breed of saunagoer identified by Megan Kressy, a longtime mobile sauna operator from Minnesota who has seen a thing or two. The Sauna Bro is easily spotted by his addiction to health podcasts, his detailed description of his sauna and cold plunge "ritual," and his certainty that he is biohacking his way to immortality (while quoting the Finnish sauna studies chapter and verse). He knows all about heat shock proteins. His sauna is hotter than yours—and his cold plunge is definitely colder.

I was a sorry excuse for a Sauna Bro. All I had was the sad gym sauna, and the more I learned, the less appealing that fetid hot box became. I was suffering from sauna FOMO, and I needed to travel to the source, the Holy Land, to experience the real thing. It was time to go to Finland.

So I rang up Eero Kilpi, a retired New York marketing executive and transplanted Finn who is the chairman of the North American Sauna Society, an organization dedicated to promoting Finnish sauna culture on this side of the Atlantic.*

A broad-faced native of Helsinki, Kilpi listened to my spiel patiently before sighing heavily into his camera. "The thing is, I've been trying to sort of, like, *avoid* talking about the 'health benefits' of sauna, for many reasons," he growled in his gravelly Finnish accent. "It's become like a sales pitch, A to B: buy a sauna, be healthier. Pardon my French, but that's *bullll*-shit. All of these people are getting it totally wrong!"

How so? I asked. It seemed like a slam dunk to me, I replied, getting all science-writer-y. I may have actually said, "What about the data?"

He let that slide. For starters, he explained, no actual Finnish person would *ever* spend just twenty minutes in the sauna, all by themselves. Maybe on a weeknight, after work. But normally, he said, Finns stay for much longer, going in and out multiple times, and usually jumping into a cold river or lake between sessions (or "rounds) in the hot sauna; there should be food, and drinks, sometimes containing alcohol, as well as long periods of just sitting around talking with friends. The entire process takes hours, not minutes, and it is more social event than health hack.

"The whole point is to *slooow downnn*," he said.

---

* Originally, the organization was called the "American Sauna Society," or A.S.S., which, based on my experiences in Finland, may be a more appropriate acronym, but the name was changed for obvious reasons.

While there probably are health benefits to sauna use, he conceded, those are likely secondary to its broader, mostly mental and social benefits, which stem from simply spending time with friends in a hot place, in nature—which Kilpi does weekly, at a friend's house on Long Island Sound.

"When I go over to my buddy's house, we spend three or four hours there, going in the sauna, or sitting around the fire pit, or going in the cold plunge," he said. "We drink a few beers, eat some sausages, and talk. And I probably shouldn't tell you this, but I always start by smoking a cigar."

That actually sounded way more fun.

Kilpi urged me to go to Finland, to experience Finnish sauna the way actual Finnish people do it. Better yet, he suggested, go in winter to see the real thing, with fewer tourists.

When I floated the idea of a winter trip to Finland to my partner, Martha, she said yes immediately, to my surprise. Her Finnish grandmother had emigrated to the United States with her family in the early 1900s, and she was curious. So in early March, we flew to Helsinki to spend ten days touring Finnish saunas, from the fanciest spas to the rowdiest ski resorts, to the funkiest, sweatiest old-school joints, like Rajaportti.

I wanted to find out: Why do the Finns love sauna so much? How do *they* do it? And is it really as magically healthy as the Sauna Bros believe?

The notion that Finnish sauna is some kind of rigorous health-optimization practice, undertaken by Instagram-ready specimens of Nordic vitality, lasted approximately one day.

On our first night, we made our way to a popular sauna/restaurant/bar by Helsinki Harbor called Löyly—the Finnish word (again) for sauna steam. Owned by the Finnish actor Jasper Pääkkönen, who played a Klansman in *BlacKkKlansman* and a Viking in *Vikings*, and is a national celebrity in Finland, Löyly is a chic, beautifully designed yet rustic space, set in a zone of redeveloped docklands. Its spacious beer decks overflow with crowds in summer, and although it is easily the most touristy sauna in all of Finland (meaning, it was the only place where we heard a conversation in English), the saunas there were also said to be very high-quality. Even Risto Elomaa, the head of the Finnish Sauna Society and an uncompromising sauna snob, had to agree.

Because Löyly was the only local sauna that was open on a Monday night, the locker rooms and common areas bustled with a mostly younger crowd, unwinding after work. The saunas are coed, so everyone wore bathing suits—anathema to sauna purists, but times are a-changing. We donned our suits, showered, took a deep breath, and walked through the sauna door.

This was our first Finnish sauna experience, and we were nervous. As expected, it was hot inside. Really hot. Way hotter than the tepid gym sauna. *Whoa, it's hot,* hot. Inside, it smelled faintly but distinctly of smoke. This was a *savusauna* or "smoke sauna," the most traditional type of Finnish sauna, and the most cherished. It has no chimney, so as the stones are heated by a wood fire, over the course of several hours in the morning, the smoke is allowed to stay in the room, before being ventilated out (and the fire extinguished) prior to people being let in. The ceiling was coated in a fine layer of ash. To the cognoscenti, the smoke is thought to add an extra note of flavor to the *löyly,* rather like peat smoke does to some single-malt Scotches. It smelled a little like an empty glass of Lagavulin. But it didn't exactly seem "healthy."

People were coming and going, and whenever someone entered, they would ask (usually in Finnish, sometimes English) whether the occupants wanted more *löyly.* More water on the rocks. Always, someone said yes. It kept getting hotter and hotter. We stayed, because we did not want to seem like weaklings in front of our hosts—Risto Elomaa, and the architect who had designed Löyly, a lovely forty-ish woman named Anu Püustinen.

It was so hot it felt like the wood benches were about to catch on fire.[*] Martha had to pull out her earrings because they were burning her ears. She had grown up with a sauna in the basement of her family's home in Michigan, but this was like nothing either of us had ever experienced. Finally, Anu made a move for the door, and we gratefully followed.

---

[*] Anu casually mentioned that this particular sauna really had caught on fire a few months earlier and had needed to be rebuilt. Someone had forgotten to monitor the stove, apparently; it happens. (Luckily, nobody was hurt.)

"Wow," Martha said.

"Wow," I answered.

Anu smiled kindly at us. There was a set of stairs leading down to a hole that had been chopped in the ice. The harbor water was murky and unappetizing, with a thin layer of ice slurry floating on the brown surface. Yet people were climbing down the stairs and dipping in. I ventured in knee-deep before realizing that this was a hard no.

The next round was easier. One thing that helped was letting go of the bogus notion, propagated by influencers, that one must sit in the sauna for a specific length of time, at a specific temperature, to achieve the mythical health benefits. Actual Finnish people stay in the sauna for precisely as long as they feel like, and then they leave. Even the Blister Boys. "You should stay in, until the idea of jumping into ice-cold water seems like the absolute best thing in the world," a sauna-savvy friend had advised. That seemed to take about eight minutes, though it was hard to tell; time seemed to slow to a standstill in the heat.

It wasn't just that the sauna was hot enough to smoke brisket. There was more to it. The scent of the smoke, the weight of the rocks, the hiss of the water flashing into steam all combined to give the heat an elemental power. The steam was not simply hot; it felt like a living thing that flowed around the room. The smell of the smoke and the wet wood imprinted itself in my mind, the way the scent of meat cooking on an open fire reaches deep into the human soul. That was not something that I had ever experienced in the sad gym sauna, with its dinky electric stove and its handful of lonely rocks, which had never known the kiss of *löyly*. That was just a hot room.

Another thing we noticed was that—how to put this delicately?—nobody at Löyly looked like they spent very much time *at* the gym, period. Or, if they did, it was certainly not the focus of their life. There was nary a Sauna Bro nor an Instagram hottie in sight. Granted, this was early March, and most of us sported a layer of winter squishiness. But nobody really seemed to care what anyone else looked like. Chalk up another win for Finnish sauna: It fosters body positivity.

Overall, it felt like a relaxed and somewhat rugged version of happy hour, only with bathing suits and without phones. We were *slooow-*

wing dowwwn, as Eero had promised. We would sleep like babies that night—yet another underrated health benefit of sauna use.

We woke up the next morning with splitting headaches. "I thought sauna was supposed to be healthy!" Martha complained as we staggered around Helsinki in a grumpy daze.

"I thought *you* were supposed to be part Finnish!"

Eventually, we figured out that we were just really dehydrated, because we had flown for seventeen hours before jumping into the hottest sauna that either one of us had ever experienced, while neglecting to drink water in between rounds. Classic newbie mistake. A big bottle of mineral water and some salty chips banished the headache and the bickering.

There are more saunas in Finland—an estimated 3.3 million—than cars. For a total population of 5.6 million, that works out to more than one sauna for every two people. They are everywhere. Most families have their own sauna at home, or at least a communal one in their apartment building, even in working-class neighborhoods. Many Finns also have saunas at their lakeside or coastal cabins, which are also quite common, not just for rich people. With about 3,000 miles of coastline, more than 50,000 lakes, and an estimated 180,000 islands, waterfront property is plentiful.

We had a sauna in every one of our hotel rooms, although we didn't use those much. There is even one in a Burger King in Helsinki. (Didn't go there either.) On cold Friday nights in winter, when almost everyone in Finland fires up their sauna, the surge in demand can cause electricity prices to spike in neighboring Estonia, just across the Baltic Sea.[1]

When we visited, in March 2024, Finland had just won the title of the "happiest country in the world," for the seventh straight year, according to something called the World Happiness Report[2]—which sounds dubious, but in fact is produced by three reputable organizations: Gallup, Oxford University, and the United Nations.[*]

I wondered: Could Finns' happiness have anything to do with their

---

[*] "Happiness" might seem tricky to measure, but the Gallup survey is quite simple: Every year, the polling firm asks about 1,000 residents of every country to rate the quality of their life, on a scale of zero to 10. Finns seemed to feel pretty good about their lot, with an average score of 7.74 out of 10—narrowly edging out Denmark and Iceland. In 2025, the Finns won again, making it eight years in a row.

national sauna habit? Or was it due to something else, like the four cups of coffee that the average Finnish adult drinks every day—also ranking number 1 in the world?

In real life, however, Helsinki was not exactly brimming with happiness. It was more than just the late-winter blues. Two years after the Russian invasion of Ukraine, lots of people still seemed spooked. More than one local that we spoke with expressed the fear that Finland might be next. Russia had bullied and occupied Finland many times in the past; that was one reason why Martha's grandmother's family had left for America. Ukrainian flags fluttered defiantly on buildings all around the city. St. Petersburg was only three hours away by train, but our spectacular new lodging, the Hotel Maria, was all but empty, in part because wealthy Russians could no longer visit Helsinki.

Which was ironic, because Finland and Russia—and Ukraine, come to think of it—are united by their love of heat bathing. The Russian and Ukrainian *banya* is a little less hot and a lot more humid and steamy than the Finnish sauna, but in every other respect, they are virtually the same: The simple wooden-bench aesthetic, the steamy heat, the sweatiness, the proximity to nature, the cold-water swimming, even the ritualistic whipping with damp green tree branches—known as *venik* in Russian, *vihta* in Finnish, and *vinyk* in Ukranian. But war had divided the world of heat bathing.

When the entire city rumbled and shook from the sounds of distant explosions one afternoon—a military exercise in the Baltic Sea, it turned out—it seemed like time to get out of town. We had heard about a spectacular new hotel, deep in Central Finland, with a spa designed by a world-renowned sauna architect. The next morning, we boarded a Finnair prop plane and flew over many miles of snow and woods and lakes to the small city of Kuopio, which, coincidentally, is where much of the influencer-hyped sauna research had been conducted.[3] We weren't there for science, however, but in pursuit of relaxation.

Perched amid a small ski resort, overlooking a frozen lake, the Panorama Landscape Hotel consists of ten cozy modular cabins scattered across a wooded mountaintop, with a top-flight restaurant presided over by a Michelin-starred chef named Kim Mustonen, who grew up not far away but had trained in top kitchens in New York and Paris. His Asian-inspired

Finnish cuisine had made the Panorama a foodie destination, but the hotel's true centerpiece is its high-design "Forest Spa," three gorgeous saunas arranged around an outdoor soaking pool, with the obligatory cold plunge and a sweeping view of the lake and mountains.

It all seemed very peaceful and, of course, healthy. We looked forward to a couple of relaxing days of sauna, skiing, and solitude. But we had unwittingly shown up on the eve of the biggest party weekend of the year, a ski race/networking event/festival of debauchery called the "Business Slalom" that originated in, and perpetuates the vibe of, the 1970s. We had no idea what was in store, as we unpacked and headed for the spa.

We had the place to ourselves for about thirty tranquil minutes, relaxing in a lovely sauna with an incredible view of the woods and the lake, then soaking in the outdoor pool, which was just about body temperature. Perfect. We exhaled deeply, finally unwinding in the peaceful Finnish forest that we had traveled so far to experience.

Then the Business Slalom people began filtering in, in groups of two, three, six, and eight. Soon all three saunas were packed with strangers sitting shoulder to shoulder and blasting out round after round of hot steam, laughing uproariously. The soaking pool also filled up, with bodies of all shapes and sizes. There was even a short line to dip in the cold plunge. Music was playing, and people were talking and flirting, sipping beers and "long drinks," a gin-lemonade concoction that was invented before the 1952 Helsinki Olympics, as part of a national campaign to steer Finns away from their overly potent vodka shots. Whatever its origins, the Long Drink is quite tasty and refreshing, even if you bring it *into* a 180-degree-F sauna, like many of the other guests were doing. At least they were hydrating.

As we watched revelers lining up at the bar, for yet another round of highly taxed beverages, Martha observed, "Everyone talks about 'sweating out your toxins'—these people want to put the toxins back *in*."

A couple of days later, still foggy from the Business Slalom, we made our way by train to Tampere, Finland's third-largest city, an old industrial town that is enjoying a renaissance as a technology center and the "sauna capital of Finland." There are more than fifty public saunas in town, in hotels, in restaurants, and by the shores of Tampere's twin lakes, but I set my sights on Rajaportti because it seemed closest to the pure roots of Finnish sauna.

Sauna bathing was already an ancient custom when Rajaportti first opened its doors in 1906. The Finns had been sweating in steamy chambers for more than three thousand years, since their ancestors arrived from somewhere out on the central Asian steppes. It may even have been old by then: The remains of twelve-thousand-year-old saunas, in the form of charred rocks, have been found in Iron Age hill forts on the Iberian Peninsula.[4]

Humans struggled to survive in the cold, so they craved warmth. The very earliest saunas were likely caves that people heated up with big fires. People who didn't have access to a cave would simply dig a hole in the earth, throw in some heated rocks, and cover it with a blanket or animal skins, and hop in—perhaps also bringing in some food to cook on

Eyewitness depiction of a Finnish sauna by the Italian traveler Giuseppe Acerbi, circa 1800. He wrote, "I once or twice tried to go in and join the assembly; but the heat was so excessive that I could not breathe."

the hot rocks, or some meat to smoke. Another plus: the smoke would have killed the omnipresent insects. As human technology progressed, the development of permanent structures, fireplaces, wood-burning stoves, and eventually chimneys would have made the heat-bathing experience more comfortable and convenient.

The undisputed kings (and queens) of heated bathing were the ancient Romans, who erected massive and elaborate public *thermae* (baths) in every major city across their empire, and the minor ones as well, from Bath in England to Lisbon to Pompeii to Tunis, and all the way east into Turkey and Syria. They had adapted the custom of hot bathing from the Greeks—the Spartans especially enjoyed a hot steam followed by a cold plunge—while adding their own sense of grandeur and luxury. The baths allowed them to show off their talents for building huge, opulent structures, and for controlling and directing the flow of water.

The Roman baths reached their apotheosis in the Thermae of Caracalla, a massive, high-vaulted complex built (by Emperor Caracalla) around 212 AD, not far from the Colosseum in Rome. Comprising 120,000 square meters (1.3 million square feet), the Baths of Caracalla were as big as Yankee Stadium, with multiple heated and cool chambers, warm and cold pools with soaring ceilings, and a coal and wood-fired underground heating system, as well as an athletic arena, a sculpture garden, even a library.

As wild as this all sounds, Caracalla was only the *second*-largest bathing complex the Romans built; the Baths of Diocletian, built a few decades later, were even bigger. Amazingly, the baths were not only intended for emperors, or even the aristocracy. Everyone was welcome. And every neighborhood and even village had its own bathing establishment, with warm rooms, steam chambers, maybe a cold room (the *frigidarium*), and typically one or more bathing pools of various temperatures. The activities were not limited to bathing, either; in larger complexes there was usually an area for playing sports or games, for cooking and serving food, and even for musical and dramatic performances.

Despite their spectacular and obviously expensive architecture, the Roman baths were cheap and often free of charge to use; common people and even slaves were admitted. "The luxurious and pleasurable world of baths afforded the greater urban populations a welcome opportunity to escape their overcrowded and cramped living conditions and the dusty streets for a few hours a day and bathe in style," writes Fikret Yegül, a scholar of classical bathing. "Moreover, for many, it was their only opportunity to bathe at all."[5]

The baths, Yegül writes, represented "the epitome of democratic ideals in the Roman society." Those ideals included access to healthcare, for bathing also served as medicine, with doctors prescribing hot, warm, or cold waters to treat specific ailments. And at least in some places and times in the Roman era, men and women bathed together, with few or no clothes.

It wasn't only Romans who bathed, either. Even the Prophet Mohammed, not known for his libertinism, urged his followers to take sweat baths, believing that they enhanced fertility. Which might not necessarily have been the case: A 2013 study done in Italy found that intensive sauna exposure over three months significantly reduced men's sperm counts and sperm function. Sauna use served as a form of male birth control, in other words. The study subjects' normal sperm production eventually returned, but only if they quit using sauna for at least six months. This makes Finnish birth rates—low as they are—seem positively heroic.[6]

On the flip side of that coin, because sauna heat tends to improve blood-vessel dilation and increase blood flow . . . rather like Viagra. Anyway, you can see where this is going: there are no definitive studies, alas, but based on what we know, it seems logical that sauna *could* help reduce erectile dysfunction. (And for those men for whom fertility is a concern, the longevity influencer and biohacker Bryan Johnson recently tweeted about his own solution to that conundrum: "icing the boys" when he sits in a sauna.)

After the Roman Empire collapsed, the Roman baths remained, and many continued to be used as such through the Middle Ages. In the Muslim world, the baths were rebranded as *hammams*, and the practice continues to this day. Sauna bathing remained prevalent in many medieval European cities as well—or so we think, because it was mentioned so often in local laws, according to Mikkel Aaland, author of *Sweat*. Some of the earliest known fire codes pertained to saunas, which makes sense: Flames plus wood minus regulation has never worked out well.

The laws protected sauna users as well as the buildings. In the 1200s, Swedish law prescribed double punishment for all crimes of violence perpetrated inside a sauna, on the reasoning that people in a sauna are generally naked and defenseless. The sauna was supposed to be sacred, ruled by the principle of "sauna peace," meaning, essentially, *don't be a jerk*—or, as a sign at an anarchist sauna that I visited in Helsinki put it, "Don't Be Stubid [*sic*]." Then, and still now, the sauna was meant to be a place where everyone could feel comfortable.

Very comfortable, in some cases. German saunas sometimes came with a brothel attached; in southern France and Italy, such establishments were known as *bordellos,* for good reason. According to Aaland, the history of sauna in Europe was punctuated by frequent crackdowns by the church and local authorities, due to the licentious and often debauched behavior that tends to flourish when people are hot, sweaty, and naked in the dark. The church eventually won out, and most if not all of the old Roman baths were closed—or repurposed as churches.[7]

Outside Scandinavia, sauna and heat bathing have been in a long-term decline ever since, until relatively recently. In places such as England and Spain, the very idea of bathing at all—as in, ever—fell out of fashion for decades, if not centuries, much to the disgust of the native peoples whom they would go on to conquer. The Spanish conquistadors were appalled to see Aztecs congregating in baths, men and women naked together, but the locals were far more repulsed by the dirty, smelly Spaniards, who had sailed all the way across the ocean without ever bothering to rinse their filthy bodies in it. In Spain at the time, thanks to a royal decree, it was customary even for aristocrats to bathe just once a year.*

By the late eighteenth century, sauna was all but unknown in Europe outside of Finland and Russia, and a few other places on the far fringes of Europe, like Estonia, just across the Baltic Sea from Helsinki. While

---

* The English Pilgrims were also reviled for their stench by the Native Americans, who they would go on to mistreat in more consequential ways.

the *hammam* survived in the eastern parts of the old Roman Empire, as well as in North Africa, in the rest of Europe as well as in North America and Australia, bathing eventually became something that was mostly done in private, in solitary bathrooms dominated by that efficient but soulless fixture, the shower.

When Rajaportti was built in the early 1900s, the surrounding neighborhood comprised some 2,500 flats or other dwelling units—only eleven of which had running water. The sauna served as a communal bathing station and water source for the entire neighborhood, as well as its news service and social network. It was once one of more than 100 public saunas in Tampere, but even as the spread of indoor plumbing and eventually tubs and showers caused most of them to shut down, being now obsolete, Rajaportti and a few others somehow survived. Rajaportti is now owned by the city, which has preserved it as-is. The guy selling tickets at the door at Rajaportti, when I visited, was serving community service for, as he put it, getting "too many fucking speeding tickets."

A traditional Finnish sauna is about as far from a Roman bath as it is possible to get, aesthetically: unfinished wood benches, a stove (usually electric now, sometimes wood-burning), some ventilation, and possibly a window. "Modern refinement such as marble pools, enamel bathtubs, showers, and nickel pipes have no place in the sauna," declares author H.J. Viherjuuri, in his treatise *Sauna: The Finnish Bath*. "The inside of the bath house must not even be oiled, timber and rough board being the best materials for it. A painted or oiled surface does not absorb the moisture from the air, and finished wood cannot allow the necessary amount of air to pass through the walls."[8]

Some architects see this minimalist aesthetic not as a deterrent, but a challenge. There are books upon books about sauna design, with gorgeous photos of simple and chic saunas wedged into all kinds of unlikely places, from shipping containers to giant barrels to floating rafts to organic, free-flowing spaces like Löyly, and making them as aesthetically pleasing as possible. Anu Puustinen, the architect, told us that she had designed Löyly to resemble the rocky outcroppings and islands that dot the 2,760-mile-long Finnish coastline, and it worked.

Even by sauna standards, Rajaportti is radically simple and old school, with no fancy architecture on display. There is no advertising or neon sign, only faded letters painted above the door that say, "SAUNA." Yet as Alex demonstrated, before opening, the sauna space was ingeniously, organically designed to create the perfect *löyly*.

Once he had heated the rocks up sufficiently to please the Blister Boys, yet not so hot as to cook casual users like me, he poured a ladleful of water into the rocks, getting it deep down into the glowing red stones (Only amateurs splash water across the tops of the rocks, apparently.) As the rocks sizzled and the steam roared out, he followed the cloud with his finger as it rose up, bounced off the ceiling, then curled back down over the upper benches where people would soon be sitting. "The person sitting there, closest to the door, gets the hottest steam," he said. "Usually they think it will be the coolest place."

Point taken.

Top left: It's Getting Hot in Here: Alex Lembke testing the *löyly* (steam) at Rajaportti sauna. (© Alex Lembke)

Bottom left: The Blister Boys take their places in the changing area at Rajaportti, a Finnish sauna that opened in 1906. (© Alex Lembke)

Right: A bather rinses his body before joining his fellow sweaters in the hot upper part of the sauna. (© Alex Lembke)

Then, as it cooled, the steam flowed down the stairs and exited via a gap under the door, while a side vent pulled in fresh air from outside. The *löyly* had a soft, almost musical quality to it, like the word "*löyly*" itself. (Again: "low-luh") "Now breathe in deep, with your nose," he instructed. The superheated air scoured my nasal passages, cleaning them out like never before. "Can you feel how fresh that is?"

He pointed to the walls, covered in a rough, breathable plaster that allowed moisture to flow in and out. The same plaster also covered the stove, which was due for its annual refurbishing (and looked it, with big cracks). Lastly, Alex pointed to the door, which was plain, narrow and short, like a Hobbit door. I had to duck to go in and out. The top of the door was just below the top of the stove, the better to retain the hot air and steam in the room. Yet the space was not uniformly hot; the steam heat had stratified into layers, so that the upper benches were at 190 degrees F or hotter, while the lower, washing-up area was a more tolerable 100 degrees F or so. But because the benches are so much higher than the stove, nobody in the sauna would ever have cold feet. From end to end, it was like a perfect machine for transmuting energy from the sun, in the form of wood, to fire to rocks to water to steam to skin, into one's very soul.

"It is perfect. It cannot be improved," Alex rhapsodized. "It is like a cathedral."

If Rajaportti is like a cathedral, then Alex is its priest. He is German but moved to Finland several years ago for his love of sauna. Like many a convert, he is more devout than most natives.

Compact, athletic, and intense, with some very interesting tattoos, he is forty-nine years old (in 2024) but looks thirty-five with his shirt off, which he attributes to his daily sauna habit. It might also be due to his love of cycling and cross-country skiing great distances across the Finnish countryside on his days off, always ending his journey at secret saunas hidden deep in the woods. But still, he seems to be a living confirmation of the Finnish sauna studies.[*]

---

[*] Ironically, he says he detests hot weather, and spends summers on the Arctic island of Svalbard, where temperatures rarely get above about 60 degrees F.

His "cathedral" still fulfills its prosaic original purpose, which was as a place to bathe. And there is no clean that is quite like sauna clean, where you feel purified down to your very pores, and deeper. Indeed, this was the original function of sauna bathing—to cleanse the body, not only of dirt, but also of the germs and other pathogens that were a leading cause of death before the development of antibiotics in the early twentieth century. Centuries before Louis Pasteur figured out that germs cause disease, the Finns and other heat-bathing cultures already knew that heat and sweating could help keep people healthy.

For most of human history, infectious disease has been our most dangerous enemy (beside other humans). We now know that high temperatures kill off most pathogens—even the Covid virus dies when exposed to sauna heat. This may also explain why, traditionally, Finnish women gave birth in the sauna; it was the most sterile room in the house, even when the men were using it to smoke reindeer meat. The old-time Finns also believed that in sauna, they were protected by *Auterinen*, spirits of healing carried in the sauna steam.

More recent studies have confirmed that the old-timers were on to something. The heat not only kills germs, but research dating back decades has found that using sauna may indeed strengthen the immune system. German studies as far back as the 1950s found that frequent sauna users got sick less often and thus missed fewer workdays than nonusers. Habitual sauna use appeared to induce what one old-school German scientist called a "hardening" or "toughening" of the immune system. But the Finns already knew that.

"The sauna is a holy place for us," says Marjo Tykkynen, a Helsinki-based photographer who I met at Rajaportti. "It's where people gave birth, died, and had sex."

But in 2025, nobody is going to Rajaportti solely to get clean (or to give birth or have sex, at least today). Most of the customers are regulars from this eclectic, hilly, bohemian neighborhood of Tampere, which is known as Pispala ("Pope's hat") because of the shape of the neighboring hill. "It's right-wing, it's left-wing—there are many architects, musicians, drug addicts, alcoholics, politicians, workers," Alex told me. "And they are all coming to this sauna, every time that it's open, and they are all sitting on the bench talking, having discussions. All these people, they are naked, and they are talking to each other. This is a very crucial function of saunas like this."

He paused.

"I think that this would not be possible in America."

I had to admit that he had a point. But maybe, I thought, it's exactly what we need.

After five rounds on the weathered benches of Rajaportti, its soft, world-renowned *löyly* had worked itself through my skin into my very soul. I was starting to get it. The heat wasn't harsh or scary at all; it was more like a gentle embrace. That did not make it easy, necessarily. The key to sauna may lie in the fact that the extreme heat is inherently stressful, physically and mentally; it takes some grit—what the Finns call *sisu*—to get through the heat, not to mention the icy plunge that sometimes follows. By stressing the body and mind with repeated bouts of heat, then cold, then more heat, sauna and similar heat bathing practices may make you stronger mentally as well as physically.

After a couple of hours in and out of the heat, I felt as though I had accomplished something. My *sisu* had gotten a workout. I rehydrated in the café with one last nonalcoholic beer, put my pants back on, and bid farewell to Alex. It was still damp and gloomy out, typical Finnish winter weather. On Christmas Day in Helsinki, the sun spends less than six hours above the horizon, and it is either rising or setting for two of those hours. As I shivered on the bus, I wondered whether sauna was perhaps a substitute, the roaring wood fire and enveloping heat the next best thing to the warmth of the sun.

I was so relaxed that when I finally got back to our hotel—which was, weirdly, located inside a hockey arena—I dozed off, waking up an hour later with the panicked realization that I'd forgotten to pay for my drinks. Not wanting to be the Ugly American, I rode the bus (free) all the way back to Rajaportti to settle up.

It was dark by the time I got there, and people were still coming and going—a family of four with two teenage boys, a pair of girlfriends, and the usual middle-aged guys sporting what Martha was now calling "sauna tummies." The older man who was holding court in the locker room was finally leaving, in a station wagon with a bumper sticker: I L♥VE 🇫🇮 SAUNA.

He was alone.

And I thought of something Alex had said: "I think there are a lot of lonesome men who come here because the sauna is the only place where they can have a physical intimacy, or close relationships," he'd said. "You could never tell them this. And if you saw them on a bus, they wouldn't say anything to anyone. But here, they are ass to ass, and it's no problem."

11

DETOXIFIED

> Sweat cleanses from the inside. It comes
> from places a shower will never reach.
> —GEORGE SHEEHAN

RETURNING from Finland, I knew that I could never set foot in the sad gym sauna again. That created a dilemma. Unfortunately, there were few saunas of any description in our city at that time, back in 2024. There was plenty of heat therapy to be found, however, but it went by a different name: hot yoga. So on a Tuesday in May, I biked over to my local hot yoga studio, just to check it out. At the very least, I thought, it might help kick-start some summer heat adaptation.

Many people assume that hot yoga is some kind of ancient Indian tradition. It's not. Bikram Choudhury basically invented the practice in California in the 1970s, claiming that the only proper way to practice Indian yoga techniques is under conditions approximating Indian weather. He heated his studios up to a withering 105 degrees F, with 40 percent humidity, and found that Americans who live in air-conditioned comfort would gladly pay to sweat like donkeys.

Before long, celebrities began flocking to his "Yoga College of India" in Beverly Hills—beginning with Shirley MacLaine and including the likes of Martin Sheen, Madonna, Lady Gaga, George Clooney,

Gwyneth Paltrow, and more. The celebrity buzz turned Bikram yoga into a global sensation, with about 1,650 franchised studios worldwide by 2006, earning Choudhury an estimated $10 million monthly.[1] For a couple of decades, Bikram yoga was literally the hottest yoga around.

Even before my mat hit the floor, I realized my terrible mistake. The room was even hotter than the KSI heat lab, with no air movement whatsoever. It made the Hotter'N Hell seem breezy. You might think twice about going outside on a day like this, let alone attempting a vigorous workout, but there were about a dozen other devotees in there, including one guy at the front of the room with NAMASTE tattooed across his abs. Even the floor felt hot to the touch. No wonder Choudhury himself referred to his studios as "torture chambers."*

This was unlike any other yoga class that I had ever experienced. Our instructor sat cross-legged on a low riser at the front of the room and began to recite a series of detailed descriptions of the poses we were to assume, signaling us to change positions with bossy little claps. This sounded like a script, and in fact it was: Every word of "The Dialogue," as it's called, was written by Choudhury himself, and it is required to be read verbatim in every class.

The Dialogue is highly detailed and uncompromising ("your forehead *must* touch your knee"), yet also manages to be cryptic:

> Bend your right arm so your elbow is touching your waist.
> Right palm is up in front of you, facing up
> As if you are asking for money from your mom:
> Give me the money.
> Do not change your palm, do not drop the money.

I mimicked the other students as best I could, kicking my right foot back to my butt and reaching back to grab the inside of my ankle. From

---

* Bikram yoga's popularity cratered after Choudhury's spectacular fall from grace. Numerous women, many of them attendees at his nine-week-long teacher-training courses, had accused him of sexual harassment and worse. The accusations were detailed in the widely viewed 2020 Netflix documentary *Bikram: Yogi, Guru, Predator* (and, before that, on a five-episode *30 for 30* podcast series: https:/30for30 podcasts.com/bikram/). Choudhury was not merely canceled, but he had fled the country, allegedly to escape paying a $7 million judgment in a lawsuit brought by a woman who accused him of raping her. But while the founder was gone, his yoga practice (and this studio) had survived.

this precarious position, balancing on one foot, we were then instructed to bend forward while kicking back and upward with our right leg, to open into "Standing Bow Pose," which is supposed to look long, elegant, and graceful.

The regular students pulled this off as deftly as ballet dancers, their toes pointed at the ceiling. My Standing Bow looked more like a Falling Tree, with limbs waving in all directions, but I managed to remain upright as sweat dribbled down my nose and puddled on the floor. I shuddered to imagine what Bikram studios must have smelled like when they were all carpeted, as the Master had once decreed.

The postures only intensified the effects of the heat; it's hard to breathe when you are doubled over with your face two inches from your knees *and* it's 105 degrees F in the room *and* your sweat refuses to evaporate because of the humidity. I thought about leaving, but there were too many sweaty, half-naked bodies in between me and the door. Then I noticed in the mirrors that a few others also seemed to be struggling. This made me feel better, because I am a terrible person.

Devotees of hot yoga claim all sorts of health benefits for the practice. Bikram insisted that the heat helps the muscles loosen up, allowing students to get deeper into the poses with less risk of injury, which seems plausible. I always felt less stiff in the hot room. Also, certain poses were supposed to affect specific body parts, improving digestion, the cardiovascular system, the lymphatic system, and so on. Overall, he often said, his yoga was meant to "purify" the body.

Suspecting that most of these claims were bullshit, I noodled around on PubMed, looking for studies of hot yoga. There were a few, mostly authored by yoga devotees and/or published in fringy, woo-woo journals. But there were some half-decent studies that did at least provide slight support for the idea that hot yoga might improve metabolic health and cardiorespiratory fitness, somewhat similar to the effects of sauna use,[2] while possibly also helping with strength, sleep, weight loss,[3] blood pressure, and so on.[4] Interestingly, some athletes had used hot yoga as a form of heat acclimation training,[5] also with positive results.

None of this was especially surprising; Bikram yoga is a demanding discipline, and any other rigorous exercise routine would likely have yielded similar physical benefits. So why would anyone put themselves through 90 minutes of sweaty, gasping exercise in, basically, the warming drawer of an oven?

"There was never a time when I didn't walk in thinking, God, I can't wait to get out of this room," said my friend Jane, a longtime yoga instructor who got her start taking Bikram classes twenty years ago, in the heyday of hot yoga. "Every class was *a slog*." Yet despite the discomfort, she kept coming back nearly every day, for five years; something about the "slog" had hooked her.

Yogis call this sort of thing *tapas*—not as in Spanish appetizers but meaning something close to "discipline" or "austerity." Appropriately, it comes from a Sanskrit word meaning "to burn," and indeed it requires a fiery discipline just to get through a hot yoga class. You really need to be locked in to balance on one foot with your arms and legs twisted around each other, in a room so hot that sweat is dripping into your eyes.*

Bikram's history and the authoritarian vibe had set off my "is this a cult?" alarm bells, but the studio was nice, and the people were friendly. Also, I had already paid for the month. I came back a few more times, and to my surprise I grew to prefer Bikram-style yoga to the "flow" classes that predominate in American yoga studios, where students are supposed to transition smoothly from one pose to the next to the next, sometimes holding each for no longer than a single breath.

Bikram poses are held longer, allowing a deeper stretch, while also requiring intense focus. I liked that we didn't need to "flow" or "float" anywhere but could concentrate on one pose at a time. Nothing forces you to be present quite like attempting a chair squat while balancing on your tippy toes with your thighs shaking ("Awkward Pose" or Utkatasana, #3 of the 26 prescribed poses).

When my introductory month ran out, I bought a pass, and became a semi-regular.

As Bikram himself had faded from the scene, hot yoga—perhaps surprisingly—had not. If anything, it had grown in popularity. Some former Bikram studios simply dropped the name and rebranded his se-

---

* The fact that Bikram himself is essentially a fugitive from justice, purportedly living in Mexico (and still offering his $12,000 nine-week teacher trainings around the world), added still another layer of *tapas*.

quence as "26+2," for the 26 postures and two prescribed breathing exercises. He had a penchant for suing rival hot-yoga purveyors, and with his downfall, many new variations on hot yoga were allowed to blossom, including heated power yoga, heated restorative yoga, and so on. New kinds of heated exercise studios also opened, offering hot Pilates (also the most popular class at my local Bikram studio) and other heated workouts.

I stumbled across one of the most extreme hot-fitness iterations in a Salt Lake City strip mall. Called HOTWORX, it is part of a national chain that offers individual yoga, Pilates, aerobics, and other exercise classes inside individual infrared sauna cubicles heated up to 130 degrees F. If this sounds like something that a bunch of fitness nuts dreamed up in a bar, you are not far off: Cofounder Stephen P. Smith, a serial fitness entrepreneur who owned a chain of spas and automated spray-tan salons called Planet Beach, was on vacation in Jamaica, drinking rum punch with business partners and friends, when inspiration struck. They were debating the best way to recover from intense exercise—infrared sauna versus yoga—when someone suggested: Why not do yoga *inside* a sauna?

The first HOTWORX opened in Louisiana in 2017, offering just that: yoga classes inside an infrared sauna. At the time, infrared saunas were the hot new wellness trend, and HOTWORX took off. It has since expanded to more than seven hundred franchised locations in the US and Canada (along with international locations in Ireland, Saudi Arabia, and Dubai). The pandemic turbocharged its growth, because the sauna cubicles allowed people to continue taking exercise classes safely and alone, outside of their homes. Evidently, there was considerable untapped demand to work out in Death Valley–like temperatures.

It sounded insane, so of course I had to try it. On a rainy Friday, I turned up at my local HOTWORX franchise—easy to find in the strip mall, with its bright-orange logo—to claim my free introductory class. A super-upbeat young woman gave a quick rundown, and had me sign a waiver, for reasons that would soon become obvious. After that, she led me back to a warren of wood-slatted cubicles. She had tried her very first HOTWORX workout just a year ago, she said enthusiastically, and liked it so much she ended up quitting her corporate job to work there full time. She was selling it hard.

From a long menu of workouts, from yoga to Pilates to cycling and even rowing, I chose a class called "Hot Buns." She handed me an orange HOTWORX-branded yoga mat and towel, pushed some buttons on a control panel, and opened the glass door for me. "How hot does it get?" I asked, stepping inside. "Up to one-hundred-and-thirty degrees," she said, shutting the door before I could protest. Not as hot as a Finnish sauna, but lots hotter than any yoga studio.

As the room warmed up, I watched ads for HOTWORX-branded supplements, as well as a warming gel that seemed quite unnecessary. Then the video class began, led by a lithe and energetic redhead flanked by two equally proficient companions. In short order, my buns were burning, along with the rest of me.

Which, according to HOTWORX, is the point. The main selling point of HOTWORX is that the heat enhances the intensity of the exercise, allowing you to get a better workout in less time. We know that heat can enhance the performance gains we get from strength and endurance training, so HOTWORX stacks them. And the workout certainly *feels* more intense, in that heat.

On my way out, the woman at the desk handed me a membership brochure urging me to join ($59 a month) and "Earn the Burn," along with a long list of the benefits of exercising in extreme heat—everything from "lowered blood pressure" to "stronger immune system" to "enhanced calorie burn." *

But the claim that really caught my eye—one of the key selling points of HOTWORX and Bikram alike—is that all that sweating helps "detoxify" the body. This notion recurs repeatedly in the fitness world, as well as in the realms of sauna and massage. It is the ultimate goal, the sine qua non of all of these modalities; all roads (and all sales pitches) seem to lead to *detoxification*. The sweatier the exercise, the more its proponents will insist that users are "sweating out toxins."

---

* Does heated exercise really burn more calories? Not clear. Exercise may *feel* more intense in the heat, but experts say you are actually doing *less* work, because the heat slows you down and creates more stress on the heart. So you sweat more, but you simply cannot do as much work in the heat as you could in a cooler environment. It's not possible. That's just how heat stress works. One could even argue that you would burn more calories in a cooler environment, because you will almost always run faster or do more work in cooler conditions than hot conditions.

Is there anything to it?

That required a deeper dive.

The idea that sweating, particularly in a sauna, somehow detoxifies the body is one of the most durable shibboleths of alternative health. It is repeated endlessly in the marketing of all kinds of saunas, as well as in pitches for hot yoga classes and other heated workout modes, and of course for spas. It's a key piece of wellness dogma.

One certainly does sweat a lot in a Bikram class, in a sauna, and definitely at HOTWORX. In that single 30-minute "Hot Buns" class, I had lost two pounds' worth of water weight—meaning I sweated out almost a full liter of fluid, in half the time it had taken me to do it in the KSI heat lab. I schvitzed so hard in that little cubicle that I felt bad for the next person who had to use it. But does all that sweat really "remove toxins" from the body, as so many seem to believe?

It at least sounds somewhat plausible. Who doesn't feel cleaner and somehow purified after a really good sweat?

According to mainstream evidence-based medicine, however, sweating absolutely, positively, 100 percent does *not* help "detoxify" your body in any way, shape, or form. Case closed. Even suggesting that it might is to invite ridicule and eye-rolling. But the actual answer is more complicated, less certain—and weirder.

While many people may have heard that sweating "removes toxins," few realize that this notion originated with the Church of Scientology, the controversial faith followed by many Hollywood celebrities, notably Tom Cruise and John Travolta. In the late 1970s, Scientology founder L. Ron Hubbard introduced the "Sweat Program," a brutal drug-treatment regimen that required recovering addicts to spend hours jogging in a rubber suit to (supposedly) sweat out remnants of illicit substances such as LSD and other drugs they had taken. This extremely arduous protocol morphed into a routine known as the "Purification Rundown," where subjects did their sweating in saunas at 190 degrees Fahrenheit, for up to four hours per day.

The Purification Rundown was initially developed to treat patients of the Scientology-affiliated drug-addiction program called Narconon. The theory was that extreme sweating would release and remove traces of drugs stored in the patients' fat tissue. In addition to the lengthy sauna sessions, patients were required to take high doses of niacin and certain minerals such as calcium and magnesium, to do light exercise, and to drink a daily half cup of "pure oils," to replace the drug-tainted fat they were supposedly losing in the sauna. In the 1980s, church-affiliated researchers had published studies purporting to show that the regimen not only released "toxins" from fat, but also lowered cholesterol, improved mood, and even increased IQ levels.[6]

The church worked hard to spread the gospel of detoxification into mainstream society. During the 1980s, a Scientology-affiliated clinic used the Purification Rundown method on oil-refinery workers in Shreveport, Louisiana, who were suffering from exposure to poisonous chemicals. That program was shut down after an independent toxicologist (hired by an insurance company) declared it to be "quackery," but the Purification Rundown popped up again after 9/11, when another church-affiliated group set up a detoxification clinic near the site of the World Trade Center. They treated hundreds of first responders there, drawing favorable coverage from the *New York Times* and other outlets.[7]

"I feel great," a Pennsylvania metalworker named Mike Wire told the *Philadelphia Inquirer* in 2007. He had come to the clinic feeling short of breath, tired, and sick all the time. He was worried he would end up like his brother, who had also worked on the wreckage pile and had died of leukemia. "I'm much healthier, more invigorated, and involved with life," he said. The group also set up a similar program in Vietnam, to help treat people who had been exposed to Agent Orange during the war.[8]

Not everyone responded well to the brutal sauna regimen. Over the years, a handful of patients have ended up in the emergency room after undergoing the Purification Rundown, including a forty-five-year-old man suffering from severe hyponatremia. In another case, the parents of a twenty-five-year-old man sued the church, claiming the Rundown caused his death from liver failure. (The church claimed his death resulted from a preexisting condition.)[9]

The Scientologists weren't the first to think about sweat this way. The ancient physician Galen believed that sweating ranked among the body's other excretory functions: Peeing, pooping, menstruating, nose blowing, and so on. And because he had observed that overweight people tended to sweat a lot when they exercised, and ultimately to lose weight, it made sense to think that sweat was at least partly composed of fat exiting the body—as L. Ron Hubbard also believed. (Turns out, it's not; eccrine sweat contains no fats whatsoever, and apocrine sweat, the kind from your armpits, contains only small amounts of lipids.[10])

For the most part, the job of cleaning and expelling toxins from the body belongs to the kidneys, along with that heroic organ known as the liver. The kidneys filter unwanted junk and waste material from our blood, while the liver metabolizes alcohol and prescription or recreational drugs that we might ingest, and just about anything else that the body recognizes as a poison. Those toxins and other waste materials typically exit in the form of urine and feces.

Eccrine sweat glands have one purpose only, and that is to cool the body by excreting water. That's pretty much it. Removing toxins and purifying the body is not in their job description. That said, a handful of studies have found that small amounts of substances such as alcohol, drugs, and even some contaminants like lead, *do* escape via sweat. This means that a person who is drunk, with alcohol in his or her bloodstream, will sweat out small amounts of alcohol (or, when hungover the next day, metabolites of alcohol). Whatever is in blood plasma or the interstitial fluid (between cells) will likely also show up in sweat.[11]

The effect is pronounced enough that some research suggests that sweat analysis and even sweat-collection patches could someday be used to test for the presence of illegal drugs—and to monitor other biomarkers as well, in a less intrusive (and possibly more accurate) way than traditional blood and urine tests.[12]

"Chemicals emerge in sweat because they happen to be floating around in blood, and the human body is inherently leaky," writes Sarah Everts in her entertaining and thorough book, *The Joy of Sweat*. "Not because your body is intentionally expunging toxins."[13] But more detailed analysis has found that sweat does indeed sometimes contain numerous other substances, including the expected sodium and potassium electrolytes, as well as trace amounts of calcium, iron, copper, zinc, magnesium, certain vitamins, plus metabolites such as lactate (produced during intense exercise), glucose, urea, bicarbonate, and ethanol (as mentioned); hormones such as cortisol, and certain cytokines (messenger chemicals including those involved in inflammation). Which is . . . not nothing. But whether these things qualify as toxins, and whether they are expelled in sufficient quantities to benefit the person in any way, is another question entirely.[14]

Nearly two decades ago, a Canadian medical-school professor named Stephen Genuis launched what he called the Blood, Urine, and Sweat Study (BUS), a concerted effort to find evidence of toxic chemicals in the aforementioned body fluids. He suspected that many of his chronically ill patients suffered from long-term exposure to environmental toxicants. His team found traces of metals such as lead and cadmium and aluminum, and chemicals such as bisphenol A (BPA), a compound used in plastic water bottles and food containers, in subjects' sweat; they also found small amounts of other unpleasant substances, including phthalates, chemicals used in making soft plastics; certain flame-retardant compounds known as polybrominated diphenyl ethers (PBDEs), which we also don't really want in our bodies; and organochlorine pesticides, a family that includes DDT. Interestingly, some of these compounds, including PBDEs, were found in sweat but did *not* show up in blood tests.

Other substances were found in much greater amounts in sweat than in blood, according to their results. Cadmium, a particularly toxic heavy metal, appeared in sweat at 87 times the concentration that it was measured in the subjects' blood. Other heavy metals also showed up in greater concentration in sweat than in urine, including aluminum (3.75-fold), cadmium (25-fold), cobalt (7-fold), and lead (17-fold). At the least, the results suggested that blood and urine testing alone may not reveal the actual concentrations of these substances in subjects' bodies.[15] (One reason may be that the liquid in sweat does not come directly from blood plasma, but from interstitial fluid, meaning the liquid outside and around our cells.)

But the BUS study only examined twenty volunteers, and its results have not been replicated by other researchers in the fifteen years since it was first published. Occam's razor points to a simpler possible explanation: That these substances were simply present in the study subjects' bloodstream, in infinitesimal amounts, and just happened to emerge in their sweat. Also, the amounts in question are so tiny that outside contamination or other laboratory error cannot be ruled out.

More recently, however, the wealthy longevity influencer Bryan

Johnson subjected himself to fifteen days of rigorous sauna sessions, testing himself (expensively) before and after. He found that consistent sauna use lowered the levels of microplastics in his blood by 87 percent, while reducing or even eliminating traces of herbicides, pthalates, perchlorate, and other industrial chemicals. He was the sole subject, so it is difficult to know whether this effect could be repeated, but it does raise questions.*

In fact, "sweating out toxins" could even backfire, for some populations—such as firefighters coming back from a fire. Firefighters are exposed to all kinds of toxic chemicals in the course of duty, from asbestos to flame retardants to smoke particles, and so on; it's a serious occupational hazard that contributes to their decreased life expectancy. If anyone needs detoxing, it's firefighters. But sweating in a sauna might not just be ineffective; some research suggests that it could backfire. Going directly from firefighting into a sauna might actually help those toxins make their way from the skin *into* the body, because the heat and the sweat tend to open up ones' pores, making skin more permeable to toxicants.[16]

Firefighters have been so interested in methods of detoxification, with good reason, that the International Association of Firefighters commissioned a definitive report on the question in 2017. After reviewing all the relevant literature, including the Scientology studies, the authors concluded that "there is insufficient medical evidence to support a recommendation for the use of saunas to remove toxicants from the body after firefighting, and the potential adverse health effects outweigh the potential benefits."[17]

Even so, I had to admit that I *did* feel somehow purified after a good sweat. I felt cleaner, both outside and in—scoured to my pores. At the most basic level, this was the original purpose of both sauna and the Roman baths: as a place to bathe, to cleanse ourselves of dirt, dust, sickness and stress. But the hot boxes of HOTWORX and the stressful routine of Bikram yoga weren't quite doing it for me, so I went in search of a true American expression of the magic of Finnish sauna. Did that even exist? What would it look like? And where has it been hiding?

---

* For a more detailed report of Johnson's sauna regimen and its measurable effects, see his X post: https://x.com/bryan_johnson/status/1997403290171330638.

12

THE CASE AGAINST COLD PLUNGING

Suffering is necessary until you realize it is unnecessary.
—ECKHART TOLLE

There was really only one place in the US where I could go to get a sauna fix: northern Minnesota. So I flew to Duluth in early May, to attend a festival called Sauna Days, an eclectic gathering held by the shore of Lake Superior.

My plan was, first of all, to spend three days doing nothing but sitting in saunas and drinking beer by the lake. But also, I wanted to check out the budding American sauna culture at its epicenter, the northern Midwestern states of Minnesota, Wisconsin, and upper Michigan. Scandinavian immigrants like Martha's Finnish forebears had gravitated to this region in the late nineteenth century, because its endless forests and chilly lakes reminded them of their homelands. The newcomers soon reestablished their beloved sauna traditions.

"Some of my earliest memories are of 'helping' my dad and grandpa build a sauna," says Justin Juntunen, a Duluth native whose family immigrated from Finland in the 1880s. Juntunen turned his childhood fascination into Cedar + Stone, a business that operates a dockside sauna on one of Duluth's old industrial piers, as well as a rooftop sauna "experience" at the Four Seasons Hotel in Minneapolis. His company also builds high-end custom saunas for clients around the world. He started making saunas in his garage in 2019, and thanks to the worldwide sauna boom he and his partner now have sixty employees and a 130,000-square-foot production facility in Superior, Wisconsin.

Sauna Days was more grassrootsy, a pop-up village of a dozen mobile and modular saunas on the grounds of a cozy lakeside resort, all run by small companies and independent operators. People wandered around in bathing suits and bathrobes, sweaty and smiling. The air smelled faintly of woodsmoke. A band was playing. I tingled with anticipation. I wanted to see how American sauna stacked up against—and maybe improved upon—its Finnish origins.

My other goal was to work on my cold-plunging technique, which I felt had fallen short, in Finland. Sauna Days featured the Mack Daddy of all cold plunges, Lake Superior: the deepest, rockiest, and coldest of the American Great Lakes. Typical water temperature in May: 42 degrees F.

I spent the first day wandering from one cozy lakeside wood-fired sauna to the next, hydrating with electrolyte drinks before switching to beer in the afternoon, as I sat by the lake cooling off between rounds and chatting with other sauna freaks. It was blissful and relaxing, maybe even better than Finland. At least Lake Superior wasn't frozen, unlike the Baltic Sea and Finnish lakes in March. The next morning, I woke up early, because I had signed up for a cold-plunging workshop scheduled for 7:30 a.m.—in the lake.

I struggle with cold plunging. Whenever possible, I dodge any occasion that might involve any part of my body (other than my mouth) encountering cold water—cold plunging, cold showering, cold swimming. This workshop, I figured, would help me face my fear once and for all. But when I woke up, it was 48 degrees F outside, with cold raindrops whapping the windows. Nope.

I snuggled under the comforter, peeking out the window as about fifteen intrepid cold plungers huddled in the water, forming a loose circle. I could hear them gasping—sorry, "doing breathwork"—even through the closed windows, as the rain sizzled into the lake, followed by a kind of group yelp. I fell back asleep knowing I had made the correct choice.

Why do I dislike cold plunging?

Mostly because I dislike cold water and pain, the dominant features of the cold-plunging experience. But I was pleased to discover that my distaste for cold plunging is at least partly justified by science.

As I dug into the literature around cold water exposure—such as it is—and talked with experts, I found that the purported physical benefits of cold plunging may have been oversold. Long touted as essential to athletes' "recovery," cold plunging in fact may do the opposite in some cases—and may even impair gains in muscle size and strength.

Also, and more to the point, most commercial cold plunges really *are* too cold—typically from the mid-forties (Fahrenheit) down to just above freezing, in some cases. The colder the better, most plungers seem to think. But research indicates most of the physiological and psychological benefits of cold exposure, such as they are, take place at temperatures in the fifties (Fahrenheit), says Mike Tipton, a professor of physiology at the University of Portsmouth who has studied cold-water exposure and survival for forty years. Below 50 degrees F or so, he says, the cold primarily stimulates the body's "nociceptors," nerves that transmit feelings of pain.[1]

All pain, no gain.

Another, more significant issue is that cold-water immersion can kill you. When the human body is suddenly dunked in cold or even cool water, says Tipton, it triggers what's called the "cold shock response." Our heart rate accelerates, while our peripheral blood vessels constrict, causing blood pressure to shoot upward. At the same time, we are flooded with stress hormones like cortisol and noradrenaline as the fight-or-flight response kicks in.

Our immediate reflex in this situation is often to suck in a big, gasping breath of air. Not good, if your face is underwater, or if a wave hits you at the wrong time. This is how many people drown: gulping for air almost immediately on contact with the cold water, says Tipton. As an ocean lifesaving expert who has worked with the British Royal Navy, he has encountered hundreds of such drowning cases in his career. The rise in popularity of cold plunging and cold swimming has brought those dangers into renewed focus.

As one astute Instagram user pointed out, replying to yet another cold-plunging post: "Gotta love it's biohacking and could be fatal at the same time."

Dedicated cold plungers see this as a feature, not a bug. The fact that we want to flee the cold water means it *must* be good for us. So we should fight off that initial urge and force ourselves to stay in, muttering, *What doesn't kill me, makes me stronger.* People who enjoy this sort of thing report feeling euphoric afterwards, with improved mood and better focus—and an insatiable need to tell everyone about how they cold plunged.*

Lastly, just personally, I don't enjoy being bullied or cajoled into anything, even things I actually like doing. Yet cold plunging seems to be almost mandatory, at many North American sauna establishments—and sometimes almost militantly enforced, by guides wielding timers and even whistles, to ensure that we realize the full health benefits on offer. During one especially weird session, Martha cowered in an icy bath for several long minutes while a sinewy hippie blew a didgeridoo in her face.

In these kinds of situations, with unpleasantly cold baths and overly officious watchers, cold plunging becomes the precise opposite of relaxing, which seems to defeat the purpose of the entire exercise. The goal of sauna bathing, remember, is to *sloowww dowwwnnn.*

"The mechanical world of objective time—seconds, minutes, hours—is irrelevant here," writes the artist, editor, and bathing connoisseur Leonard Koren in his cult classic book *Undesigning the Bath.* "Taking a bath properly requires being able to guiltlessly linger, hang out and/or do nothing whatsoever."

---

*As a side note, not to name names, but some of the public cold plunges that I've visited have seemed a little bit, shall we say, *janky*, especially if multiple previous users have neglected to shower after their sweaty sauna sessions. Cold plunging is difficult enough without having to do it in the sweat soup of dozens of random strangers. So do your part to prevent jankification, and shower or rinse off beforehand.

A strictly timed two-minute ice bath torture kills that vibe, turning it from guiltless lingering into a "protocol."

Can't we just, uh, chill?

Cold-water bathing has a long and somewhat checkered history as a health hack. The ancient Greeks and Romans used deliberate cold-water exposure to treat various complaints, notably fever, which makes sense; so did the ancient Egyptians. The "frigidarium"—a cold-water pool or cold room—was a standard feature of Roman baths, alongside the steam rooms and warm pools. Thomas Jefferson soaked his feet in cold water every day, supposedly to maintain his health, and maybe it worked: He lived to be eighty-three, more than double the average life expectancy for his time. Swimming in icy water also seemed to help the old naked gents at the Finnish Sauna Society.

In the nineteenth century, wealthy patients paid good money to be subjected to "hydrotherapy" (getting blasted by jets of frigid water), at elite health spas like the Battle Creek Sanitarium in Michigan, overseen by Dr. John Harvey Kellogg, who also invented corn flakes cereal. Most interestingly, some eighteenth-century European mental institutions employed a tactic called the *bain de surprise*, or "bath of surprise," suddenly dunking their patients in cold water, with no warning, to jolt them out of their depression or psychosis.[2]

As we'll see, this might actually be the *most* science-based use of the cold plunge.

More recently, professional athletes and their coaches have taken up cold-water bathing, believing that it helps with recovery and reduces soreness after a tough workout or competition. LeBron James was an early ice-bather, as was Olympic swimmer Michael Phelps. Photos of pained-looking athletes sitting in icy tubs have been a social-media staple for years. Tour de France cyclists and other endurance athletes have experimented with full-length "ice boots" that cool the legs by circulating chilled water around them. And of course, most adults are familiar with the age-old "RICE" protocol for dealing with a sports injury: Rest, Ice, Compression, Elevation. It seemed obvious that ice and cooling are precisely what athletes' tired, sore, overheated body parts needed.

There is some logic to it: Cold exposure does dampen inflammation, for example. Also, some of the longest-lived creatures on Earth spend their lives in cool or cold environments, such as the Greenland shark, which lives as long as five hundred years—and basically cold plunges for its entire life.

Then there is the Wim Hof factor. Beginning around 2000, the eccentric Dutchman became globally famous for his feats of endurance in freezing water. He has sat in an ice bath for nearly two hours, which was a record at the time (since broken by another crazy person), and has also set a record for long-distance swimming underneath ice, which sounds like, and was, a really bad idea. (His corneas froze over during the 188-foot swim, making him blind and unable to see the exit hole.) Nevertheless, thousands of acolytes have paid to study the "Wim Hof Method," a combination of intense breathing exercises, repeated cold-water (and cold air) exposure, and meditation techniques designed to teach people how to withstand extreme cold.

Hof has done more than anyone else to popularize cold-water immersion as a fundamental wellness practice for everyone, not just athletes. The internet is full of videos of him and his followers jumping into ice holes and hiking up snowy mountains in their Speedos. What is less well known is that more than a handful of Hof's followers have died while using his methods, including some who were possibly overwhelmed by the cold shock response.

A 2024 investigation by the *Sunday Times* (London) uncovered eleven cases in the UK alone where people had died while using Hof's methods or similar techniques.[3] "They drowned because they were hyperventilating in water," says journalist Scott Carney, who helped popularize Hof with his 2017 book *What Doesn't Kill Us,* but has since become critic of Hof's program (although he still cold plunges almost daily at his home in Colorado). Carney believes the worldwide death toll of Wim Hof followers could be as high as thirty-three.[4] (Wim Hof's website and other platforms now include this warning: "Engaging in these exercises near water increases the risk of shallow water blackout, which can lead to drowning and, tragically, death.")

The second way that cold water can kill you is by causing cardiac arrest. The initial shock of entry, and the ensuing cardiac strain, can trigger a quick heart attack—especially in older men with preexisting and/or undiagnosed heart conditions. This possibility crosses my mind every single time I swim in cold water, as a fiftysomething male.

And if your face is *also* submerged, that stimulates the vagus nerve, which wants to slow the heart down. This creates what Tipton calls "autonomic conflict," as the sympathetic (fight-or-flight) and parasympathetic (calming) nervous systems battle for control of the heartbeat. This in turn can lead to arrhythmia, an irregular and uncontrolled heartbeat that can be dangerous by itself, and particularly so if one is also trying to stay afloat in cold water at the same time. Tipton's research has found that even young and healthy people have about a 1 to 3 percent chance of going into arrhythmia, on entering icy water.[5] (Although not all arrythmias are fatal or even serious, he adds.)

Tipton began his career studying why and how people died in cold water, after falling from ships into rough and freezing seas; he is a worldwide authority on cold-water lifesaving, and a longtime consultant for the British Royal Navy. Only in the last decade or so has he had to contemplate what happens to people who seek out frigid waters *voluntarily*. It's been quite a journey, and he remains puzzled by the popularity of cold plunging.

"I'm not the Fun Police, but amongst all the activities you could do, plunging yourself into cold water is at the high-risk end of the spectrum," he sighs. "So there's even more reason for making sure that you are a person who *should* do it—and that when you do, it's done safely."

Is the risk (not to mention the discomfort) worth the supposed rewards?

One case where cold-water immersion is *always* helpful, even lifesaving, is when it's used to cool victims of heatstroke, who stand a far better chance of survival if they are immersed in cold water immediately—as we saw in chapter 8. KSI director Douglas Casa knows this firsthand: As a teenager in the 1980s, he suffered a severe heatstroke while competing in a state high school championship 10K race on a hot and humid June day. A quick-thinking athletic trainer saved his life by dunking him in ice water—which was not standard practice at the time. That incident, he says, launched him on his career as an evangelist for heat safety.

Beyond that, however, there is more hype around cold plunging than solid data. One extremely popular claim by cold-plunge partisans, for instance, is that cold exposure activates our "brown fat," a special type of fat tissue that burns energy (unlike regular "white" fat) to generate heat. This fat is said to convey almost magical health benefits, including reducing the risk of diabetes and other chronic diseases, as well as improving our cold tolerance. Unfortunately, most adults typically have only a few grams of brown fat (unlike infants and small children, who have quite a bit).* So any beneficial effect from activating brown fat is likely also quite small. Even a study of Wim Hof himself, using fMRI and other imaging techniques, found that his brown fat activation (after a session of the Wim Hof Method) was "unremarkable."[6]

Multiple studies suggest that cold-water immersion (the scientific term for cold plunging) *may* help reduce muscle soreness after exercising. LeBron was on the right track. But it might also be counterproductive in other ways: Some studies have shown that cold-water immersion immediately after resistance exercise could reduce or limit the growth of muscle and improvements in muscle strength, post-workout. In other words, dunking yourself in cold water may wipe out any gains from the strength workout that you just did.[7]

One 2015 study found that cold-water-immersion after resistance training reduced subsequent muscle growth by 20 percent.[8] In another, far more elaborate study, scientists had subjects perform a single intense resistance-training session, focused on the legs and lower body. They then placed the volunteers in a strange contraption that immersed one leg in very cold water (46 degrees F/8 degrees C) and the other leg in warmish water (86 degrees F/30 degrees C). While in this device, the subjects consumed a special protein drink that enabled the researchers to observe how much muscle protein synthesis was taking place in each leg. They found that the cold leg had less blood flow to the muscles, and less uptake of amino acids from the protein drink, and thus likely less muscle growth, than the warm leg.[9]

This and other research suggests that the hot part of "contrast therapy"—the sauna or hot tub—may be more helpful for muscle health and exercise gains than the cold plunging part. One study of Finnish basketball players found that they recovered their strength and explosive power more quickly if they ended their workout with an infrared sauna session, rather than passive recovery at room temperature.[10] Another interesting and well-run trial had volunteers perform

---

* Infants and young children typically have a significant amount of brown fat, but we lose nearly all of that brown fat by adulthood, which is why claims that cold water is doing something magical to our health via brown fat activation need to be taken with a grain of salt.

more than an hour of "exhaustive" arm cycling intervals (think stationary bike but using handheld cranks instead of pedals). They recovered far better from this ordeal when their arms were warmed post-exercise, than when their muscles were cooled.

Further investigation using mouse muscle fibers demonstrated that at lower temperatures, muscles are less able to "refuel" (via a process called glycogen resynthesis) than at warmer temperatures.[11] In yet another study, researchers found that cold-water immersion impeded healing, in subjects with laboratory-induced muscle damage, while warm water speeded healing *and* reduced soreness.[12] Lastly, heat therapy—not cold—has been found to mitigate the muscle atrophy that comes with disuse due to injury or immobilization in, say, a cast or a boot.[13]

It all makes sense mechanistically: Warm water (or a warm sauna) opens blood vessels, increasing blood flow to the tired or injured muscles. It also activates heat shock proteins, whose job is to repair damaged cells. Cold water, on the other hand, does the opposite—constricting blood flow, blunting repair mechanisms like heat shock proteins, and also making muscles and connective tissue less elastic. This is why we "warm up" before exercising, to get those muscles nice and soft and flexible; nobody cools down *before* they exercise.

Even the scientist who came up with RICE in the 1970s—rest, ice, compression, and elevation, to recover from injury—has recanted the "ice" part of the protocol. "Coaches have used my 'RICE' guideline for decades," confessed sports physician and bestselling author Gabe Mirkin, in a 2015 blog post, "but now it appears that both Ice and complete Rest may delay healing, instead of helping." (Ice is still useful for reducing swelling in cases of injury, he added.)[14]

His reasoning was interesting: Cold tamps down inflammation, which cold-plunging partisans claim as one of the benefits of cold plunging. But in the context of injury, lowering inflammation sometimes turns out to be a bad thing. While it was long thought that inflammation was purely harmful, newer research has found that it also acts as a signaling mechanism that helps marshal the body's own healing response—and that it also helps spur the strength and endurance improvements brought on by exercise. Cold water or ice blunts inflammation, and thus slows or stops that healing response, as well as adaptation to the stress of exercise. (Worth noting: Cold-water immersion does not appear to dampen training-induced improvements in aerobic fitness—only strength.)[15]

Cold-water immersion or cold plunging in between sauna sessions may have another, hidden downside, points out Earric Lee: The cold may dampen or even negate the beneficial effects of the sauna heat. Because the cold water brings sauna users' core temperature back down, he explains, they may never get hot enough inside to activate a robust heat-shock-protein response, or the beneficial cardiovascular changes that come with heat adaptation. "Your core temperature needs to rise up to a certain level before a lot of these chemical processes in the body can happen," he says. Cold water stops that in its tracks.

On the other hand, in the long-term Finnish sauna studies we talked about earlier, those study subjects were presumably doing cold dips in between their sauna sessions—in Finland, it's the done thing.

Of course, many if not most of the studies cited above—both pro and con cold-water exposure—were quite small, with twenty subjects or fewer, the vast majority of whom were healthy, fit males in their twenties (i.e., graduate students volunteering as research subjects for beer money). Relatively few subjects were female, or older than forty. Also, these studies tended to be short-term, sometimes involving as little as a single ice-bath or hot-water-immersion session. And for obvious reasons, it is impossible to do a truly blinded study of cold exposure, which means that there is a high possibility of a placebo effect tainting the results, especially in studies with subjective endpoints—for example, where subjects are asked to evaluate their mood or pain level.

This also appears to be the case with studies of cold water and mental health. A randomized controlled trial published in 2024 found that a Wim Hof–based protocol involving hypoxic breathing exercises and cold showers reduced depressive symptoms by 24 percent, anxiety by 27 percent, and stress by 20 percent in a group of forty-one depressed women—a significant reduction across the board.[16] But the control group, who did slow breathing exercises with warm showers, *also* reduced their symptoms by the same amount. Only a handful of other studies have been done using the Wim Hof Method, looking at physiological and psychological endpoints, and those were largely inconclusive, apart from a modest effect on inflammation.[17]

Studies of non-athletes and cold water immersion have been even more ambiguous. One recent review looked at twenty-four studies of cold-water exposure in nonathletes, dating back to the 1950s, and determined that "the existing literature presents conflicting outcomes and, in some instances, inconclusive findings." Also, the reviewers found a "high risk of bias" in fifteen of the twenty-four studies.[18]

I was able to find only one truly large-scale cold-water study, but it proved to be fascinating. In the Netherlands, researchers assigned more than three thousand people to take cold showers every morning. These folks ranged in age from eighteen to sixty-five years old, and they were randomized to end their usual daily shower with 30, 60, or 90 seconds of cold water every day for a month. Incredibly, they actually did it. The cold showerers missed about 30 percent fewer workdays than a control group who took only hot showers every day. This was especially interesting because both groups reported the same number of total illness days—but for whatever reason, the cold showerers felt more motivated to go into work.

This may speak to why people swear by cold plunging and cold showering, with an almost religious fervor; die-hard cold plungers insist it gives them more energy. And it may even be slightly addictive, for some. In the Dutch cold-showering study, many of the subjects voluntarily continued with the cold showers, after the initial 30-day study period expired. Even though they didn't have to (and, presumably, were no longer being compensated for participating in the study). So they clearly liked it on some level.[19]

Indeed, people are flocking to cold plunging and cold-water swimming, from hyper-driven CEO types to middle-aged and older women. In Pembrokeshire, England, a small local group of female cold-water swimmers calling themselves the "Bluetits" started out with just twenty-five members in 2016, bravely taking to the chilly waters despite everyone else in the 'shire thinking that they were insane. When governments closed swimming pools and other indoor recreation spaces due to the Covid pandemic, the Bluetits' membership exploded to more than 150,000 swimmers of both sexes in 159 chapters, or "flocks," spread all around Britain's chilly 7,700-mile coastline.

This points to perhaps the most consistent reported benefit of cold-water exposure, and that is its effect on mood and mental health. People do it because, for some reason, it makes them feel better. A study of more than seven hundred mostly female Scottish cold-water swimmers, aged eighteen to sixty-five-plus, found that nearly all of them at least believed that it improved their psychological well-being.[20] But the electrifying thrill of dunking oneself in cold water, says Tipton, is the very same thing that makes it dangerous for some people.

"We've argued that if there is a benefit, it's from the cold shock response, which is [also] the most dangerous response evoked on immersion because it's the one that leads to heart attacks and drowning," Tipton says. "However, that sudden fall in skin temperature releases quite a lot of stress hormones, and ends up releasing serotonin. So you get a feel-good factor, it activates you. It's the thing that makes you feel alive."

Fair enough. Back at Sauna Days, the cold rain stopped, and the sun came out. The lake looked a lot more inviting, especially after a few hot sauna sessions, although I knew that the water was exactly as cold as it had been in the morning. Even so, a little bit of FOMO kicked in. Other people were frolicking in the lake and having lots of fun. *Why not me?*

I made my way to the rocky lakeshore and ran into Glenn Auerbach, the founder of Sauna Days and one of the prime movers behind the revival of sauna in North America. His shaggy graying hair and ultra-laidback, hippyish demeanor—not to mention his penchant for wearing bathrobes—reminds one of The Dude from *The Big Lebowski*, but he is a man of purpose. He had fallen in love with sauna while hitchhiking around Finland after college and came back to his native Minnesota as a kind of missionary for sauna culture. Unfortunately, this was the 1980s, when sauna was still unfairly associated in the public mind with gay sex and AIDS. It's been an uphill battle until the last few years. Auerbach had launched Sauna Days in 2019—just in time for Covid. But this was actually good timing: Since the end of the pandemic, sauna has exploded in popularity worldwide, perhaps as a reaction to the forced isolation that we all endured. And Auerbach had come into his own, publishing a blog called *Sauna Times*, as well as hosting a podcast (often recorded in hot saunas).

"Sauna is life-changing, people really groove on that," he'd said, when we first spoke.

The Dude was about to swim in the lake, so I tagged along. At the rocky shore, I gingerly waded in and squatted down in the frigid water, making sure to keep my hands and, crucially, my nipples out of the water, a cold-plunge cheat code; it feels twice as cold with one's sensitive bits immersed. Instead of pain, I felt only relief, as all that pent-up sauna heat dissipated into the chilly lake water. I lowered myself neck-deep and let out a deep, satisfying sigh.

A couple of other festivalgoers splashed in the lake nearby. Nobody seemed bothered by the cold water in the slightest, nor were they watching the clock, gritting their teeth, or self-seriously blowing didgeridoos in each other's faces.

At that moment, I realized what I had gotten wrong about cold plunging. My mistake was viewing it through the same lens as the biohacking health "optimizers," as a purely physical practice wrapped in bro science. That wasn't it, at all. Maybe it was more about changing your mental state, knocking you out of whatever thought-loop or spiral you're stuck in—rather like a modern-day *bain de surprise*.

The second mistake had been to think of cold plunging as a purely solo activity, a torturous ordeal to be endured alone. It was much less painful in the calming company of The Dude, I found. In the lake, with the other sauna nuts, it became less about the pain and more about bond-building with others who were going through the same thing. And I had to admit, there's a certain thrill to splashing around in water that chilly. It felt a little naughty. It was no longer about *tolerating* the cold water, which implies a focus on suffering; rather, we were *accepting* it, the way we must accept all the suffering that is inherent to life. Doing it together made that acceptance easier.

The water is just as cold for everyone else, I thought, *just like life itself*—

Just then, my Buddhist reverie was shattered by two sweaty, meaty, red-faced guys who tumbled into the water right beside us, laughing and whooping and splashing. Their bodies were still steaming from the sauna, and I think maybe they'd had a couple of beers.

"I love it, man!" Glenn said, with a huge smile. "Just a bunch of hardy dudes!"

After Sauna Days, I tried to approach cold plunging with a more open mind. Sometimes, it felt like just the right thing, the perfect way to rebalance—not to mention thermoregulate, my new favorite word. Other times, I was just not in the mood.

But on the good days, I came to appreciate that cold-water bathing really was doing something useful. Like sauna, it helps crack open our thermal window, expanding the narrow temperature band in which most of us live our daily lives—our "thermostatic environment," as Mike Tipton puts it. Just as sauna tests our thermoregulatory system, cold water immersion does the same, at the opposite end of the spectrum. If one is good for you, then shouldn't the other be, too? Isn't the point to get outside of the Goldilocks zone? In either direction?

Remember, our ancient ancestors had to deal with both heat *and* cold. "In evolutionary terms, we used to get these thermal perturbations normally," Tipton told me. "The body does want to maintain a fairly constant internal environment, in terms of metabolism and biochemistry and temperature and all these other things like oxygen levels. However, to do that, it's got all these mechanisms that it's developed over the years. And my belief is that, unless you perturb them, you don't keep them functional in the same way."

In other words, he said, the hot-cold contrast is like exercise for our thermoregulatory system. "We don't question the importance of 'use it or lose it,' in terms of exercising the musculoskeletal system," Tipton said. "But perhaps that applies to thermoregulation as well? So, like with exercise, we're now having to reproduce something artificially that once happened naturally."

Also, even I had to admit that cold plunging in a heat wave—or just a normal 95-degree F Salt Lake City summer day—is the absolute best.

It all finally made sense when I traveled to Tampa, to visit a unique hybrid yoga studio and sauna space called Kodawari. The name means "pursuit of perfection" in Japanese, but neither the studio nor its owner, a short, fiery woman named Annette Scott, quite fit the stereotype of the Type A, perfectionistic yogi. Tucked in between a Dollar Tree store and a headbanging CrossFit studio, on the southeast side of town, Kodawari offers guided sauna and cold plunge sessions that focus on building social connection—through heat and cold. It's not about health optimization, but about another side effect of sauna that I couldn't help but notice: *feeling better.*

I had heard that Scott had a somewhat unique take on hot and cold. Coming out of a long career as a yoga teacher, she practiced what she called a "bio-psycho-social" approach to sauna, with no talk of "biomarkers" or "toxins" allowed. And she had some rather unique insights into why someone might benefit from a good sauna and cold-plunge session, delivered in a decidedly non-yoga-like vocabulary.

"This might increase your metabolism, and maybe you'll get better sleep, and it's possible that you'll take a great poop the next day," she told me. *A great poop?* "I can talk about how your immune function is going through the roof, and I can tell you how your lungs and cardiovascular output are going to be positively impacted," she continued, "but what I'm really doing is helping you find a way to be less of a cocksucker."

Oh.

"Not that there's anything wrong with that," she hastened to add. "But it's really about developing character and resilience."

To that end, one Wednesday morning, a dozen of us trooped into a very hot sauna for a one-hour session. This wasn't supposed to be silent, solo exercise, where we would all sit there marinating in our sweat and thinking about how we were improving our biomarkers. Instead, we were supposed to share, and talk. We began with a "check-in." How were we feeling? What had brought us here—to a sauna in Florida in June?

I was a little surprised when the elegant, obviously affluent blonde mom sitting next to me, with perfect makeup and gold earrings, piped up with, "I'm in here so I don't *stab* someone today."

Everyone else looked at her stunned, like, even *you* feel that way?

We trooped outside to the requisite cold plunge—four repurposed hot tubs arranged on the parking lot, next to a dumpster—and slipped in. I was a little worried to get into the same tub as Blonde Mom, but a couple of other people also joined us, and she did not appear to be carrying a knife, so it felt safe. We were all in it together, literally. And on a hot humid June morning in Tampa, the cold water felt delicious.

As she settled into the cold water, her mood visibly lightened, and she let out a long sigh, closed her eyes, and said, "Don't you just *love* feeling sauna drunk?"

I did. I did love feeling "sauna drunk," although I had never thought of it in those terms. But I knew the feeling, sort of light-headed and detached from worldly problems, almost like an out-of-body experience. Pretty much the opposite of wanting to stab someone.

The Japanese, who rival Finns in their love of heated bathing, have a word for this state: *totonou,* which literally means "tidied up" or "in order," but in this context its meaning might be closer to "in tune" or perhaps just "super chilled-out." Another way to think of it is as like a post-sauna trance. The *totonou* typically happens in the resting phase, after the hot sauna and the cold-water dip. It is hard to measure or quantify, although one small Japanese study had found that subjects in *totonou* had altered brain waves.[21]

I practiced saying it out loud, in a guttural Cookie Monster voice: *"To-to-know."*

I hadn't expected that the mind-body highway would travel through heat, or cold. But apparently it does.

And all of a sudden, it made sense. All of those good feelings that came from unplugging from the world, as one does in a sauna—from *slooowwwwing dowwwwn,* as Eero Kilpi had put it—that was the whole point. It was all about getting Sauna Drunk. *Totonou.*

Did it have any basis in biology, or was it just a mood shift? A placebo-driven illusion? Was it the heat shock proteins doing their thing, or was it something simpler? The sense of accomplishment for enduring the discomfort? The feeling of connection, from going through it all with other people? The yin and the yang of the hot and the cold?

To find out, I traveled to Colorado to take part in a highly unusual sauna study designed to answer that very question.

13

THE HEAT CURE

> In the midst of winter, I found there was,
> within me, an invincible summer.
>
> —ALBERT CAMUS

I'm lying on a hospital bed, drenched in sweat and burning up from the inside. I want it to stop.

Outside, it is a lovely summer day in Vail, Colorado, with tourists whizzing around on electric bicycles or heading off on relaxing mountain hikes. In this sterile, fluorescent-lit room, machines beep and staffers hover anxiously about, watching me suffer. A kind woman named Ashley Mason presses chilled cloths and ice cubes to my forehead. *You're doing great*, she murmurs. *You're going to make it.*

There is no hope of escape, because I am encased in a green metal cylinder that looks like part of a rocket fuselage but is in fact a personal-sized infrared sauna. My head is exposed, but the rest of me is tucked inside this tube, which is lined with heating elements that are slow-roasting me with 1,000 watts of power.

This is by far the least fun "sauna" that I have ever experienced.

But it's not really meant to be fun. This is research, which means that once again I have agreed to stick a temperature probe where probes are stuck. (It's like I never learn.) The upside, if one can call it that, is that I am unlikely to die, because my core temperature is being carefully monitored by said probe. Presumably, were I to overheat, pass out, or start convulsing, somebody would do something.

I volunteered to be a test subject for a novel clinical trial co-designed by Mason, the kind woman with the ice cubes, who is a professor of psychology at the University of California at San Francisco. The goal is to test the effects of "whole-body hyperthermia"—basically, getting people really hot—on symptoms of severe depression.[1] It's a novel idea that occupies the cutting edge of psychological research, albeit with less hype than studies of psychedelic drugs like ketamine and psilocybin. Mason and her colleagues hope that this study might ultimately help validate heat treatment as a legitimate mental-health protocol that could give hope to millions of suffering people.

Lying in the tube, I have doubts. Data and common sense tell us that heat waves and lack of access to air-conditioning are strongly associated with emotional disturbances, short tempers, and an increase in violent behavior. So, pretty much the opposite of better mental health. How could getting hot help people feel less depressed?

Yet this crazy-sounding theory is supported by some compelling evidence. For some unknown reason, depressed people seem to struggle in the heat—even more than the rest of us. Physicians and researchers have long observed that patients with major depression don't sweat very much, and sometimes not at all.[2] At the same time, these folks also seem to maintain a higher-than-normal resting body temperature, most of the time.[3]

This curious symptom had been widely noted in passing, mostly in smaller studies, but Mason had helped to confirm it in a huge, twenty-thousand-subject dataset of healthcare workers compiled during Covid.[4] The primary purpose of that study, called TemPredict, was to monitor the workers via their Oura sleep-tracking rings, to see whether changes in their body temperature could predict when they might be coming down with Covid. But when Mason plotted the subjects' 24-hour body temperature data against their self-reported mental-health status, the pattern was unmistakable (the upper line represents the most severely depressed subjects):

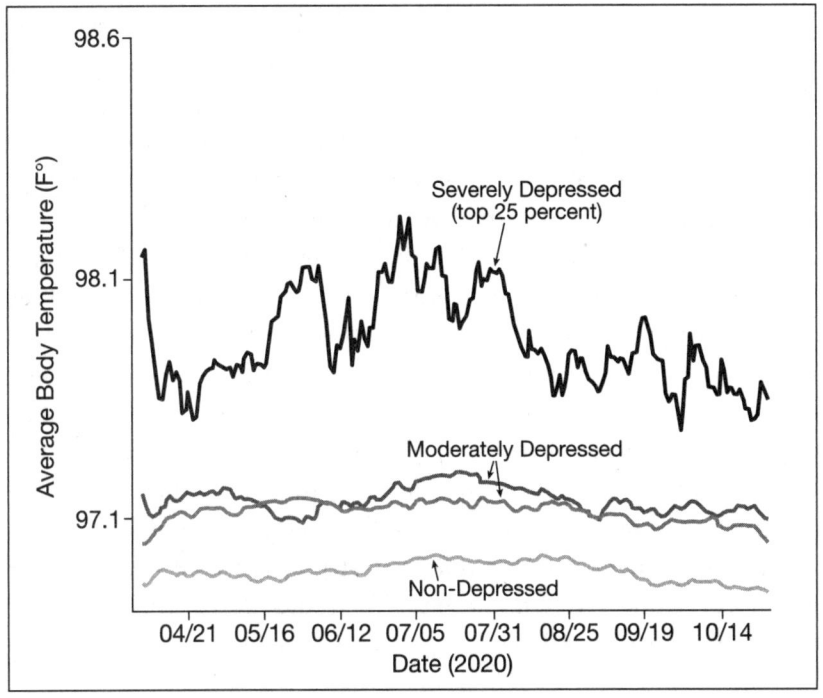

A 20,000-person study found that severely depressed people maintained a significantly higher body temperature than moderately or non-depressed people. (Mason AE, et al. 2024)

The link could not have been clearer: Severely depressed people run hotter, for some reason. They maintain a slightly but distinctly higher core body temperature than non-depressed or even moderately depressed people. It semed to be similar to the temperature dysregulation experienced by menopausal women as hot flashes—only the "hot flash" had persisted for six months. "It's nice when you don't need statistics to see what is going on," Mason says.

But what did it mean?

That part is trickier. Depression is a very common disorder, but because of the way it is diagnosed, based on checklists of symptoms but with no single defining feature, it manifests in a multitude of ways. "Depression is different for everyone," says Mason. "You could be sleeping all the time, with no interest in eating or activities—or you could be not sleeping at all and eating all the time."

But this temperature anomaly appeared consistently across the board in severely depressed people. It was the one symptom that many shared in common.

Even more intriguing, when very depressed people received treatment, in the form of drugs or electroshock therapy, their body temperature would revert to normal, some studies found.[4] And their ability to sweat also returned. (This is similar to what happens with heat adaptation: one's resting body temperature drops slightly while sweat rate increases.) In fact, one side effect of some modern antidepressant medications is hyperhidrosis, or excessive sweating.* This is so common that physicians often prescribe a second medication to control it, called Ditropan (initially developed to help with bladder control). In depressed people, it appeared that inability to regulate their body temperature was somehow tied to an inability to shake off sadness.

"Our hypothesis is that people with depression have a harder time thermoregulating," Mason told me. "Their 'thermostat' is broken. The treatment gives it a workout."

Does it ever.

I am only the second person to try out this grueling hot-tube protocol; when I visited in June 2024, the study was just getting started. The first test patient was Vail Behavioral Health executive director Chris Lindley, who told me that it was "one of the hardest things I've ever done."

I later learned that Lindley wakes up every morning at 5 a.m. to do high-intensity interval workouts, followed by a plunge into ice-cold water. Oh, and while serving with a medical detachment in Iraq, his unit had survived a suicide bombing. Yet *this* was the hardest thing he's ever done.

---

* Older, tricyclic antidepressants actually decreased sweating in patients. Interestingly, another common antidepressant called bupropion, used as a second-line therapy and also to aid in smoking cessation, was found to improve athletic performance in the heat: In one study, athletes could maintain a higher power output and higher core body temperature, without feeling hotter. Watson P, Hasegawa H, Roelands B, et al. 2005. "Acute dopamine/noradrenaline reuptake inhibition enhances human exercise performance in warm, but not temperate conditions." J Physiol 565 (Pt 3): 873–883. https://doi.org/10.1113/jphysiol.2004.079202. https://www.ncbi.nlm.nih.gov/pubmed/15831540.

On the other hand, Lindley noted, he did feel better after the treatment. "I'm a happy, optimistic person, and even I noticed an effect," he said. "It was like looking through polarized glasses. Everything was crisper and more alive."

This study is called "CHILL'D" (for "Cold and Heat to Investigate Lowered Levels of Depression"), because at some point, I'll be required to submerge myself in a tub of 49-degree-F water that is sitting just a few tantalizing feet away. As far as anyone knows, this is the first-ever trial combining heat treatment *and* a cold plunge for depression. It grows out of decades of scattered data, from sauna studies and small random experiments, suggesting that targeted heat exposure—and cold—may have some significant mental-health benefits.

This is not a new idea. One of the most curious but unexplained findings of the long-term Finnish sauna studies was that frequent sauna users—most of whom presumably cold-dipped in between sauna rounds—experienced fewer psychotic episodes over their lifetimes.[5]

But nobody had ever really investigated the possible effect of sauna on mental health. Most research has focused on how heat affects the physical body. Yet some or maybe most of the widely discussed health benefits of sauna, like the reduction in heart attacks and strokes observed in those long-term Finnish sauna studies, may in fact originate in the mind, rather than from mechanistic improvements to our cardiovascular system.

According to Earric Lee, who participated in some of the Finnish sauna research, at least some of the observed cardiovascular benefits may even originate in the nervous system, rather than in mechanistic improvements to our cardiovascular machinery. In particular, he says, sitting in a sauna appears to induce "sympathetic withdrawal," meaning it pulls back or quiets the sympathetic, fight-or-flight side of our nervous system—letting the parasympathetic, or "rest-and-digest" side, take control. That, in turn, calms us down, slowing our heart rate, lowering our blood pressure, possibly reducing our risk of heart problems—but also, making us feel better, my favorite "health benefit" of sauna.

As noted in the previous chapter, there is at least some suggestion that cold plunging itself could have a beneficial effect on symptoms of depression, even temporarily. Studies have found that dunking ourselves in cold water substantially boosts our blood levels of both dopamine and noradrenaline, a neurotransmitter that increases alertness, cognitive function, and blood pressure.[6] Both chemicals are produced in the body's fight-or-flight response, in response to the sudden stress of cold. Yet while cold-plunging studies have found elevated levels of those chemicals, as well as norepinephrine, in subjects' blood post-plunge, that does not mean that they are also elevated in the brain (thanks to the blood-brain barrier).

Even so, the fact that cold-water-immersion does lower inflammation, boosts serotonin and beta-endorphins (the body's natural opioids) in the blood, and stimulates the vagus nerve, suggests that cold-water exposure may have some beneficial effect on depression.[7]

But nobody had really dug in to figure out if heat plus cold could be turned into a legitimate mental-health treatment, as opposed to an Instagram biohack.

Now it is being put to the test, across the street from some of the most expensive ski lifts in the world.

Eagle County, Colorado, is home, or second home—and, in some cases, third or fourth home—to some of the richest people in America. Their multimillion-dollar chalets cluster around the ski resorts of Vail and Beaver Creek. It is a beautiful place in summer and winter. How could anyone feel depressed here?

But Eagle County also has a substantial working class, folks who live in modest apartments and ramshackle trailer parks in "down-valley" towns you've never heard of like Gypsum and Dotsero. (In a modern American mountain community, the richer you are, the higher the elevation at which you can afford to live.) These are the worker bees who make the resort economy run, the food servers and dishwashers and bus drivers and hotel cleaners and landscapers. They are overworked, underpaid, essential and yet, too often, treated as virtually invisible.

For a variety of reasons—including wealth inequality, an adrenaline-seeking extreme sports culture, the nonstop partying of resort life, the plethora of grueling service jobs, the high elevation, and other factors that are harder to pin down—mountain towns like this can be hotbeds of substance abuse, domestic violence, and mental health problems, among other pathologies. In 2016, Eagle County was rocked by sixteen suicides in a single year, three times as many as it typically had. The following year, another seventeen residents killed themselves, representing a wide range of social backgrounds. During 2018, someone in the county would attempt suicide on 324 out of 365 days. At the time, the entire county population of fifty-five thousand people was served only by a handful of private psychiatrists and psychologists, most working on their own or in small practices. Only three mental-health providers accepted insurance.*

In the fall of 2017, local voters passed a ballot initiative to fund community mental-health services with a $60 million tax on cannabis. Since then, Vail Behavioral Health has raised another $92 million from wealthy donors to establish a network of clinics in the valley, and to extend coverage to working people and resort employees through a program called MountainStrong. The organization now employs thirty mental-health service providers, including five psychiatrists, and is building a new twenty-eight-bed inpatient hospital, half for adults and half for adolescents. In addition, Vail Behavioral Health is also funding and hosting research studies like this one, looking at alternative mental-health treatments.

Depression is their primary target, because it is one of the most prevalent medical conditions, period, yet also among the trickiest to treat. Nearly three in ten American adults have been diagnosed with some type of depression, at some point in their lives. It can be so subtle and well-hidden that it is hard to spot, but it is far from a static or harmless disease. In the US, 1.6 million people attempted suicide in 2022, according to data from the US Centers for Disease Control and Prevention (CDC). "Only" about fifty thousand succeeded, but that still means that suicide ranks well ahead of automobile accidents as a cause of death in the United States.[8]

All of this occurs despite the fact that depression and related disorders are already very widely treated, typically with SSRI (selective serotonin reuptake inhibitor) medications like Zoloft, Prozac, and Celexa.

---

* The story is told in the 2023 documentary film *Paradise Paradox*, produced by Olympic skier Bode Miller.

In England, antidepressant prescriptions climbed 35 percent in six years, with almost 15 percent of the population taking at least one antidepressant drug during 2022.[9] In the US, one in nine adults (and one in four women) took some form of antidepressant in 2023, according to the CDC—and something like 50 percent of those patients are on more than one medication at a time. Yet suicide rates have climbed steadily, rising 37 percent (in the US) between 2000 and 2022. White men are especially proficient, comprising nearly 70 percent of successful suicides—while also ranking among the least likely to seek treatment for mental-health issues.

Around the world, it is a similar story. SSRI drug usage has climbed sharply over the last three decades, growing by multiples of 2x to 10x in the UK, Canada, Czechia, Spain, Finland, France, Germany, Italy, and Korea.[10]

But the standard treatments for depression have been falling short. While SSRI antidepressant drugs like Prozac have completely changed the field of psychiatry, undeniably helping many patients, they definitely don't work in everyone. Data shows that for every five people who receive an antidepressant, one is helped. Even so, SSRIs had performed so well in their FDA-mandated clinical trials that they were a slam dunk for approval, beginning with Prozac (fluoxetine) in 1992.

But when skeptical scientists later reviewed *all* of the early clinical trials of SSRI drugs such as Prozac—including some studies that had remained unpublished when the drugs were first approved—it turned out that overall, the placebos had actually performed better than the drugs. The placebo effect is particularly strong in mental-health studies, with between 20 and 40 percent of patients reporting improvement in their symptoms in some cases simply because they *believe* they are getting effective treatment.[11] Even so, it appeared that SSRIs had been oversold.

Even as antidepressant prescriptions grew exponentially over the last three decades, the prevalence and toll of depression have also mounted around the globe. SSRIs may be an effective treatment for some, but clearly, they have not proved to be the "cure" for depression that many had hoped they would be.

Which is why Vail Behavioral Health is so interested in new and innovative approaches to mental health treatment, including not only this CHILL'D study but a separate study looking at the use of psilocybin and other psychedelics—an ambitious agenda for a small-town mental health clinic.

"I think there are better ways to deliver this care," says Chris Lindley. "The current model is really two methods, talk therapy and medication, but too many people have *not* had their health improved by those two methods. For some people it's great, but not for everybody."

His goal, he says, is to try to validate new tools that can be used to augment existing treatments—beginning with heat. "My agenda is to bring these services to the masses a lot faster," he said. This study will combine the heat treatment and cold plunge—so-called "contrast therapy"—with cognitive talk-therapy follow-up. They're hoping to recruit up to 100 patients with significant depression to see, in effect, whether heat can help replace Prozac.

There is already some data suggesting that it could do just that, including some from a very unexpected source: hot yoga.

While noodling around on PubMed, early in my Hot Yoga Era, I had stumbled across a striking study of Bikram yoga and depression, by a group based at Harvard. Behind a mouthful of a title—"A Randomized Controlled Trial of Community-Delivered Heated Hatha Yoga for Moderate-to-Severe Depression"[12]—lay a revelatory finding. The researchers had put thirty-three depressed subjects through eight weeks of Bikram-style hot yoga classes, and at first look, the study appeared to be a bust: Only two-thirds (twenty-two) of the subjects had even made it through all eight weeks, and those who did had only averaged a little more than one class per week, when they were supposed to attend two. Not promising, but understandable.

Yet in the end, even a single weekly hot yoga class had appeared to have a major effect on their depression: 16 of the 22 people who completed the program showed at least a 50 percent decrease in their depressive symptoms, on standard clinical evaluations—and 12 of those 16 had experienced a total remission, meaning they no longer reported any symptoms of depression at all. In other words, their depression had effectively been cured, by hot yoga.*

This seemed like a pretty big deal. In this small study, hot yoga had vastly outperformed antidepressant drugs including SSRIs, which tend to work for only about 20 to 30 percent of patients, with side effects including weight gain and loss of libido. These yogis saw their depression mostly or even completely relieved, at least temporarily. The primary side effects: Improved strength, balance, heat tolerance, and flexibility. (And of course, sweating.)

There had been a handful of other studies of plain old room-temperature yoga and mental health, but those had not found yoga to be any more effective than either antidepressant drugs or exercise (which has also been found, reliably, to reduce symptoms of depression).[13] So this Harvard study appeared to hint that the heat itself may have worked some sort of magic against depression.

As I dug deeper into the literature on heat and mental health, I discovered a faint but steady trickle of studies that supported the idea that heat exposure could affect mood in positive ways. One researcher's name kept popping up: Dr. Charles Raison, a maverick psychiatrist who has been exploring the boundaries between mental-health science and Eastern spiritual practices for more than thirty years.

As a young doctor, Raison became fascinated by a mysterious sect of Tibetan Buddhist monks who practiced a form of yoga and meditation called *Tum-mo* (or *Tummo*). Tummo is a very intense form of breath-based meditation and visualization exercises whose goal is to generate heat inside the body, known as the "inner fire." The details are complex, and it takes years if not decades to master the technique, but essentially, with each long, slow, deep breath, the monks would visualize heat accumulating inside the "vase" of their body. Somehow, it worked.

---

* A control group, who were only put on a waitlist for hot yoga classes, showed only a 6 percent symptom reduction, likely due to a placebo effect. It might have been more revealing to compare one group of subjects doing heated yoga versus another group doing the same yoga routine at room temperature, but nobody asked me. So it is possible that the reduction in depressive symptoms may have had at least something to do with the social interaction and sense of purpose that would come from attending regular yoga classes.

These monks lived almost as hermits, high in the mountains, and their "final examination" was uniquely arduous, according to Alexandra David-Néel, an intrepid Frenchwoman who had lived in Tibet and studied Buddhism in the early twentieth century: "The neophytes sit on the ground, cross-legged and naked," on a moonlit, windy night, she wrote in her 1932 book, *Magic and Mystery in Tibet*. "Sheets are dipped in the icy water, each man wraps himself in one of them and must dry it on his body. As soon as the sheet has become dry it is again dipped in the water and placed on the novice's body to be dried as before. The operation goes on in that way until daybreak. Then he who has dried the most sheets on his body is declared the winner of the competition."[14]

Which sounds like something that your weird neighbor who is "into Wim Hof" might attempt. But it was apparently real, if perhaps overly dramatized. Back in the early 1980s, a Harvard Medical School professor named Herbert Benson had visited a Tummo monastery in the Himalayas, hoping to document the phenomenon. As head of the brand-new Mind-Body Institute at Harvard, Benson had done pioneering studies of meditation and its effects on physical parameters like blood pressure and heart rate.[15] His 1975 book, *The Relaxation Response*, was a huge bestseller and helped popularize meditation in middle-class America.[16]

Benson persuaded three veteran *Tummo* monks to let him stick temperature sensors all over their bodies, including their navel, spine, armpits, arms, and fingers, and of course fitting them with the requisite rectal probes. Then they meditated. As their meditation progressed, their rectal temperatures did not rise, but their bellies got warmer by about 2 degrees C. Their fingers and toes heated up even more, by as much as 8.3 degrees C, or about 13 degrees F. There appeared to be something to the *Tummo* legends.*

The Harvard scientists further poked and prodded the perplexed

---

* There is some dispute about whether the monks actually dried the sheets on their bodies, as reported by David-Néel. At the monastery, Benson learned that earlier reports had lost something in translation: The monks did not actually *dry wet sheets* on their bodies, but rather, they *wore dry sheets*, and only sheets, in extremely cold weather. Not quite the same, but nonetheless impressive.

monks, measuring their body fat with calipers, sticking electrodes on their heads to measure brain waves, and collecting their exhalations via breathing masks to measure their oxygen consumption, as if assessing the fitness of elite athletes. But none of these measurements really seemed to explain how the monks generated so much heat in their extremities.[17]

When he read Benson's 1982 report in *Nature*, Raison told me, "I got obsessed, fascinated by this question: What does generating body heat have to do with Enlightenment?"

He decided to devote his career to unlocking the secrets of Tummo. He and a Buddhist colleague at Emory University even attempted the wet-sheets trick themselves, on a "cold-ass" morning, and found it as difficult as it seemed. Hoping to study the mysterious, snow-melting monks in person, he courted the head of the leading Tummo monastery in exile, which is located in India. But the monks seemed reluctant. When the abbot visited him in Atlanta, Raison treated him to a fancy steak dinner to try to make his case. "One of the great moments in my life was walking into Fogo de Chao with two fully robed Buddhist monks," he told me. "But he made it clear to me that these guys did not want to be studied."

As for the connection between temperature and Enlightenment, he would have to figure that out for himself.

As a traditionally trained medical psychiatrist, Raison had prescribed antidepressants to hundreds of patients and had taken them himself at times. But like many of his colleagues, he had become disillusioned with their performance. So he began looking for alternatives.

One potentially promising avenue had to do with inflammation, which appeared to play a powerful role in mental health, just as it does in physical health. Depressed people seemed to have higher levels of pro-inflammatory cytokines, or chemical messengers, in their brains.[18] These chemicals are typically released as part of the immune response, marshalling the body's resources to fight off infection and kill potential pathogens. In a certain subset of depressed people, this immune-based inflammatory response was switched on all or most of the time, possibly causing some of their symptoms of depression.

But certain interventions appeared to reduce or relieve this brain-based inflammation, notably exercise. When we exercise, we temporarily increase levels of certain inflammatory chemicals, including one called interleukin-6 or IL-6.[19] But that activation is transient; in the long run, post-exercise inflammation levels end up lower than before we started. This is because exercise induces a classic stress response, where a minor and temporary discomfort (exercising) leads to a longer-term benefit or improvement. We strain our muscles, but they end up getting stronger. We induce inflammation, but it ends up lower. This inflammation-lowering effect may help explain why exercise has been found to be effective against symptoms of depression.

As it turned out, the answer he was looking for came not from exercise, but from the study of infection and fever. While Raison was investigating the role of inflammation in depression, he happened to meet a University of Colorado neuroscientist named Christopher Lowry, who excitedly told him a story about depression and fever. Lowry and others had been studying an African soil microbe called *m. Vaccae*, which induces a mild but harmless fever in humans. Researchers had found that controlled *m. Vaccae* infection made leprosy vaccines work better in patients in areas of Uganda where the disease was still endemic. Along the way, they also discovered that people who had been infected with this microbe reported improved mood and reduced anxiety afterward.

Digging deeper, Lowry found that infecting rats with *m. Vaccae*, and the ensuing fever, appeared to activate a specific brain region called the dorsal raphe nucleus, which is responsible for regulating mood, anxiety, and feelings of reward.[20] This is the exact same brain region that is targeted by antidepressant drugs, because it is largely responsible for producing the neurotransmitter serotonin. Then, Lowry stumbled across an old paper in the University Colorado library suggesting that simply heating up laboratory animals could have the same effect, causing this part of the brain to release serotonin. He and his colleagues repeated the old experiment and found that heat did indeed appear to activate some of the same brain mechanisms as SSRI antidepressants.

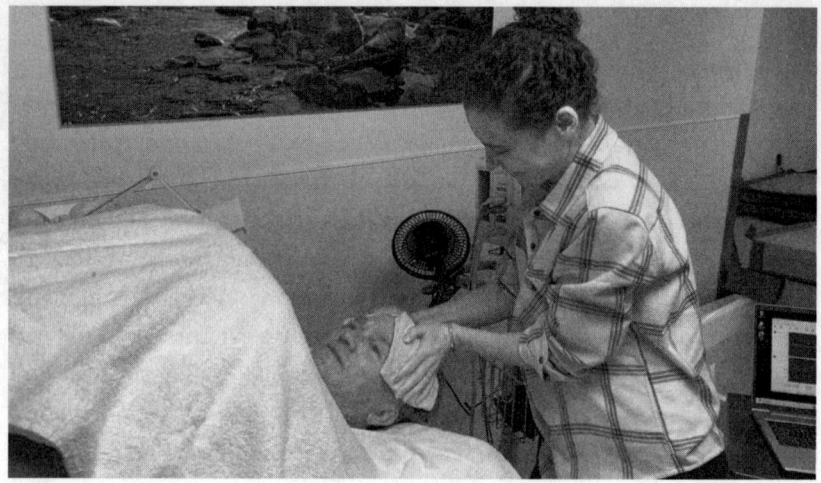

Medium Rare: The author lies in the infrared sauna, midway through the heat treatment, as lead investigator Ashley Mason applies cool cloths to his red, sweaty face. (Courtesy of the author)

It raised the question: What if, instead of antidepressant drugs, they simply gave depressed patients a fever?

Which is why I'm lying in the Hot Tube: They're trying to give me a fever.

Or at least mimic one, by heating my core temperature above 101.3 degrees F (38.5 degrees C)—the level at which physiological changes linked to heat adaptation begin to occur, such as heat shock protein activation. Unfortunately, it's taking a while, as my body struggles to maintain homeostasis. The towel under my back is drenched with sweat, and it requires all my *sisu,* my sauna grit, not to simply quit. If Mason stops with the cold compresses, if she pauses even for half a minute, I've decided, I'm out.

"You might want to scooch the probe in a little farther," Raison suggests, hovering nearby. My eyes widen in alarm.

"That might make his temp go down," warns Chloe Sorensen, the research assistant who has been shepherding me through the treatment.

Whew.

Tall and lean, dressed in khaki shorts and a red plaid shirt, the sixty-seven-year-old Raison (who goes by "Chuck") is understandably nervous. This study, which he and Mason designed together, represents a make-or-break moment for decades' worth of research, going all the way back to the Tummo monks.

In 2016, Raison had published a small but novel study in *JAMA Psychiatry* that was a precursor to this one (and a sequel to the heated-rats study): He and his colleagues had put sixteen severely depressed people into an infrared chamber and heated them up to a core body temperature of 101.3 degrees F (or 38.5 degrees C), the magic temperature threshold where heat adaptation begins.[21] The study was both randomized and controlled; a control group of subjects had been placed in the device and merely warmed, not heated. Also, importantly, it was double-blinded, meaning the researchers themselves did not know which subjects had gotten the treatment and which had not.

The results were dramatic: After just a single heat treatment, study subjects shed more than half of their depressive symptoms. And when the researchers followed up, six weeks later—expecting the effects to have worn off by then—they found that the subjects were *still* much less depressed than they had been. Heat treatment appeared to be powerful and long-lasting. The only patients for whom the treatment had not worked at all were the three patients who had been taking SSRI drugs.

Raison started to think that he might be onto something.

The device he used, known as a Heckel HT3000, had been invented in Germany to aid in cancer treatment. In the 1930s, European researchers had discovered that heating up their patients mimicked the body's own native healing mechanism of fever, and it seemed to weaken cancer cells.[22] Which is less crazy than it sounds: Back in the late nineteenth century, a New York City physician named William Coley had observed that one of his supposedly terminal cancer patients had been mysteriously cured after going through a serious bacterial infection and the ensuing severe fever.[23]

Coley is now recognized as a pioneer of immunotherapy, treatments that harness the immune system to attack cancer, but he was also fascinated by the possible uses of heat and fever in healing. Ancient medical traditions, including Chinese medicine as well as Ayurveda, have long prescribed heating (and cooling) the body to treat a wide range of ailments, from arthritis to cancer. The ancient Greeks would sometimes intentionally give patients fever to cure them. Hippocrates deployed heat against inflammatory illnesses and even cancer. (He also advocated treating fever with the bark and leaves of the willow, which ultimately led to the modern wonder drug known as aspirin.) In 1927, an Austrian scientist named Julius Wagner-Jauregg had received the Nobel Prize for using a mild malarial fever to cure patients of paralysis brought on by syphilis.

The Heckel and predecessor devices had been used in Germany, as an adjunct to chemotherapy. When patients were heated up in the machine, their chemotherapy treatments appeared to work better, likely because the heat softened up the cancer cells' membranes, making it easier for the medicine to penetrate and kill them. The drawback was that it required getting patients so hot—with core body temperatures higher than 104 degrees F (40 degrees C), for well over an hour—that many of them needed sedation merely to endure it.[24] More recently doctors have devised safer ways to deliver heat directly to cancer tumors, such as with focused ultrasound, which has shown promise against certain specific cancers such as glioblastoma (a type of brain tumor).[25]

Raison had found the Heckel device sitting unused in the basement of a German hospital. After his JAMA paper, he had it shipped back to the US to use in further mental-health studies. But the US Food and Drug Administration decided that the Heckel posed "significant risk" to study subjects and demanded additional safety studies before it could be used. Raison didn't feel like dealing with the red tape, so he gave up.

Mason picked up the baton (or perhaps, the rectal probe) and ran with it. She spent most of her time helping patients with sleep disorders, but her first interest is heat. She frequents saunas and steam baths like San Francisco's famed Archimedes Banya, a magnet for Bay Area tech leaders and intellectuals. She and her partner always felt better after a sauna, and after reading Raison's 2016 study, she decided to try to understand why.

The 2016 study was not a one-off. A different group of researchers published a subsequent study in 2020, comparing exercise and hot-water immersion for symptoms of depression, and found that subjects who spent 15 to 20 minutes in a 104-degree-F (40-degree-C) hot tub, twice a week, lowered their depression scores significantly—and by much more than those who had simply followed an exercise program. Also, the hot-tub patients were far less likely to drop out of the study than the exercisers. So it seemed that heat or hyperthermia held at least some promise as a therapy for depression.[26]

The goal of the current study is to build on her and Raison's laboratory findings, and possibly scale them up into a treatment protocol that might actually help real people. In place of the cumbersome, costly, regulatorily challenged Heckel, Mason had found this Clearlight Dome infrared sauna that I'm currently lying in, which does roughly the same thing as the Heckel for a fraction of the price (about $2000) and none of the red tape.

"I love saunas, I love steam rooms, you name it," she told me. "And I do think there's mood effects that I notice—I feel great. But I also think the data are so compelling, and nobody else is going to do this. So it might as well be me."

Which is how we all ended up in Vail.

It doesn't take a PhD in the molecular biology of the brain to understand, intuitively, that sensations of warmth might engender positive feelings and states of mind. It is built into our language, the way we talk about people and relationships: "Warm" people are friendly and helpful, while "cold" individuals are hostile and ungiving. We sign letters with "warm regards." Our friends and family, we hope, will receive us "warmly" when we visit. We connect with our partner in the "heat of passion."

These are not, it turns out, merely figures of speech.

In the mid-2000s, a PhD candidate in psychology at Yale named Lawrence Williams[*] devised a clever experiment to see if these personality traits were linked to actual temperature sensations. In the experiment, some forty undergraduate subjects were told that they

---

[*] Now a professor of marketing at the Leeds School of Business at the University of Colorado at Boulder.

were going to interview someone for a job, and then give their evaluation of the person. But the true point of the experiment was cleverly hidden.[27]

Each subject was escorted to the interview room, located on the fourth floor, by a research staffer. The key to the study was what happened during the elevator ride: On the way up, the staffer would ask the subject to hold a cup of coffee—either hot or iced—while they pretended to write some notes. The interviewee on the fourth floor would behave identically with each subject, following a prepared script. Yet the study subjects came away with vastly different impressions of the person, depending on whether they had held a warm coffee cup in the elevator, or a cold one. Those who had held the warm coffee tended to rate their interviewee much more generously than those who had held the iced coffee.

Other researchers have found similar connections between physical warmth and social warmth or happiness. In a study from UCLA, two psychologists placed volunteers in a functional MRI, which measures brain activity in real time.[28] While in the MRI, the subjects then read messages from family and friends, while holding onto either a warm pack or a room-temperature ball. The subjects who held the warm pack felt more connected to and loved by their friends and family than those holding the cooler ball. The researchers concluded, "These results suggest that a common neural mechanism underlies physical and social warmth."

But what might that mechanism be?

Nobody really knows. In part it may have to do with memory, a primal connection of warmth and love. Even the ancients had observed that temperature and mental health may be linked. Greek and Roman physicians believed that autumn and winter chill could bring on depression. Going much farther back, our ancestors also knew that being cold could easily kill them, and that warmth equaled safety.

As the authors of the coffee-cup study wrote, perhaps "feelings of warmth when one holds a hot cup of coffee or takes a warm bath might activate memories of other feelings associated with warmth (trust and comfort), because of early experiences with caretakers who provided warmth, shelter, safety, and nourishment."[29]

But it may also have to do with a phenomenon called *interoception*, which means our awareness or perception of internal body sensations—in other words, how in touch we are with our own inner body, our gut, our organs, our blood, our skin. It represents, literally, the body-mind-brain connection. Some researchers believe that this is how physical states can affect our mental health and wellbeing. The warm coffee cup elicits a warmer emotional response via interoception. In depressed people, however, that connection may be broken—hence their elevated body temperature and impaired sweating response. They have lost touch with their internal environment.[30]

One theory behind the heat treatment, says Christopher Lowry, a molecular biologist at the University of Colorado and a member of the research team, is that simulating a fever, then dunking patients in freezing-cold water, could break through and re-establish that interoceptive connection between body and brain, and shock patients out of their depression—just like the *bain de surprise*, from hundreds of years ago.

"Depressed people have an overactivity of something called the default mode network, [which] is on when we're thinking introspective thoughts," Lowry says. "It could be guilt, shame, regret, or grief in depressed people—just thinking about themselves. But if you engage with the environment, meaning you have a social interaction, or you're knitting, or walking outside or reading a book, that system shuts down and other systems turn on."

"And so we thought, there's certain types of sensory stimulation that if you *don't* shift your attention from introspection to engagement with the environment, you will die. If you get too hot and you don't seek water, you don't seek shade, you don't find a way to cool off, you'll die. And so that kind of stimulus enforces a shift from default mode network, depression- like functioning, to actually doing something to solve the problem."

As I stare up at the hospital's dropped ceiling and fluorescent lights, trying not to think about my mounting discomfort, I start to wonder, *how depressed do you have to be to make this seem appealing?*

Pretty depressed. This study is only open to patients who qualify as "severely depressed," according to a widely used clinical depression "instrument" known as Beck's Depression Inventory.

What does that mean? Many people have experienced simple depression—temporary periods of sadness or low mood that may have affected their relationships or their work lives. Life events such as losing a job, a breakup, or the death of a loved one can kick us into clinical depression for weeks or months, even a year.

Severe depression, or major depressive disorder, is an order of magnitude worse. You don't just feel a little sad, you feel utterly empty and hopeless. You lose all interest in doing things that once brought you great joy. You move slower than normal, talk slower, and think slower. You feel fatigued to the point of being physically wiped out. You sleep terribly at night, and you can't concentrate during the day. And you feel worthless, like you are the cause of all the world's problems, an Everest of guilt piled on your head, with no hope of relief.

Officially, I am here in Colorado purely as an observer—and also, I suspect, so Mason and Raison can be sure their treatment protocol does not harm any actual study subjects.

But my interest in this study is more than strictly professional. As I read about the symptoms and sequelae of depression, a sense of recognition began to sink in. The symptoms sounded a bit too familiar. I had been feeling a little bit down, myself, for various reasons. A couple of weeks before driving to Vail, I had taken Beck's Depression Inventory myself. It asks users to rate their feelings on a scale of zero to three, across twenty different questions. Zero is fine, three is worst. The results confirmed that I had a solid case of moderate depression. In subtle but specific ways, that mid-grade depression had harmed my work life, my income, and my happiness—which, being depressed, was something that I didn't think I deserved. And I had done nothing about it.

Which is the real reason why I had volunteered to undergo the arduous heat treatment myself—a decision that I've come to regret, now that I am wilting like an order of French fries left under the heat lamp at Hooters. But maybe it will help break me out of my funk.

At long last, Chloe announces, "You're done!"

The meat thermometer lodged in my rear has reached the magical temperature of 38.5 degrees C (101.3 degrees F). She switches off the sauna, and I "rest" for another two minutes, like a tenderloin roast, before she slides open the tube, exposing my sweaty torso to the magically cool air. It's a huge, indescribable relief. I want to weep. But when I try to sit up too quickly, my head swims. I am way past "sauna drunk," more like dollar-shots-in-Cabo drunk.

Guided by unseen hands, I shuffle over to a white porcelain tub that looks like it belongs in a luxury hotel bathroom. The water is clear and inviting. I haven't looked forward to a cold plunge this much ever. I step in carefully, one foot and then the other, and then drop my butt right down into the water, as I've learned to do—you can't hesitate.

This is what I've been dreaming of, for at least the last hour. Nice and cold, as advertised. But it's actually *too* cold; weirdly, my skin is burning. Chloe is there with the stopwatch, ticking off seconds, and it takes so long for her to say, "You've been in a minute and a half."

I take deep breaths, and clasp my palms together, Namaste-style, for thirty more long seconds until she calls time.

Still groggy, I shuffle across the hall to rinse off in the shower and get dressed. I am too sauna-drunk to drive, so I ride back with Raison and Mason, who pepper me with questions. What was it like? How do I feel?

I don't know, beyond feeling really glad that it's over. I do not feel notably less depressed, in the immediate aftermath—but I do feel a lot more physically wrecked. I need food—large amounts of really unhealthy, satisfying, salty foods. We stop at the first restaurant we see, where I inhale a burger with sweet potato fries and an ill-advised beer. I go to bed early, only to lie awake with a crushing migraine. After a warm midnight shower, some salty snacks, and three ibuprofen tabs washed down with Gatorade, I finally fall asleep in the wee hours, feeling skeptical about the whole thing. Have I just wasted my time? And my sweat?

My home in Salt Lake City is about six to seven hours by car from Vail. To break up the trip, I stop at Glenwood Springs, Colorado, a classic old pool and hot springs complex dating to the late nineteenth century. Long ago, people from all over the country would travel there by train to breathe the pure mountain air and bathe in the warm, faintly sulfurous pools. That hasn't changed; I float around for an hour or so, among the Midwestern families and the older ladies out for their afternoon swims.

It is calming and almost magical.

As I float, I ruminate on the Treatment. Was it really worthwhile? I don't see how this heat treatment is going to help anyone; I'm mostly glad that it's over. Do I even *like* heat anymore? The Glenwood pools feel more like my speed—and some other studies have suggested that plain old hot-water bathing, like in a bathtub, can have its own antidepressant effects.[31]*

Back on the road, something strange happens. Driving is one of my least favorite activities, because it is so deadly boring, yet deadly dangerous. Even on a trip to the drugstore, there is a nonzero chance that you could be carelessly killed because someone decided to read a text from their dog sitter at the wrong moment. I'm always on edge.

But this time is different. Past Grand Junction, dark clouds begin massing to the south—the summer afternoon monsoon, common in the Southwest in June and July. This is a pretty big one. Soon rain showers are dropping from the gray cloudbanks like curtains, causing massive flash floods (I'll later learn) with waterfalls pouring into the canyons around Moab, Utah, just to the south. The wind picks up, whipping tumbleweeds across the roadway and pushing my car toward the shoulder as I drive. Rain bands drum on the windshield, over and over. Trucks are swerving.

Normally, I would be freaking out in a situation like this. But I am focused, floating on a feeling of pure liquid calm. I am *supposed* to be driving through this storm, flowing smoothly through the curves. I set the cruise control to 85 miles per hour, and let the highway draw me home, across the desert and then up and over the mountains, toward rays of sunlight beaming through Biblical clouds.

---

* There are other, easier ways to get one's body temperature up, anyway—such as Bikram yoga. A University of Wisconsin study of twenty experienced Bikram yogis found that after a 90-minute class, seven of them attained core body temperatures higher than 103 degrees F, and one male practitioner got all the way up to 104.1 degrees F, well above the 101.3-degree-F level where heat adaptation begins.

I feel zero anxiety whatsoever. I turn the radio on and remember that tonight is the first presidential debate of the 2024 election. I listen to two minutes of grumpy old men arguing about golf, then switch it off to ride home in untroubled silence. I get home in what feels like no time.

14

HOTWIRED

> What should young people do with their lives today? Many things, obviously. But the most daring thing is to create stable communities in which the terrible disease of loneliness can be cured.
>
> —KURT VONNEGUT

For the next few weeks, the world looks just a little different. Sharper, as Chris Lindley had promised. I feel stronger and more resilient—life's challenges no longer seem so overwhelming, and other people don't annoy me quite as much. I stop dwelling on grievances and regrets. And I begin looking forward to the future.

I take the Beck's Depression Inventory test again and score an 11—half my previous tally.

Did the hot-tube treatment do this? Is it some kind of placebo effect? Or a bit of both?

"People say they feel a sense of mastery after they've done it," Raison tells me later. "Depressed people often feel like: I can't do anything, I'm a failure, I'm hopeless."

That tracks. I feel glad to *have done* it, kind of like how I'd felt after a Bikram class, or the Hotter'N Hell Hundred. After that, I can do anything. For the next few weeks, I am calmer and more productive than I had been in months.

My session in the hot tube capped off a year and a half of exploring the power of heat. I was now heat-adapted. Heat-educated. And most importantly, heat-*resistant*. I had been reborn as an athlete. My sweat glands flowed like Niagara Falls. I was always optimally hydrated. My heat shock proteins were firing. I could handle even the hottest sauna session, with withering blasts of *löyly*. Even my climate anxiety was in slight remission. I had found the cure for the summertime blues. And my biomarkers suggested that I was likely to live forever.

Well, maybe not quite—the jury is still out on that. But I am a completely different person from the guy who had shown up at UConn, terrified of a hot day and unable to insert a rectal probe correctly. I love to sweat now. I love going for hot bike rides on empty trails; I can fearlessly book a midday tee time, on the sultriest August day, and have the golf course pretty much to myself. I might even sign up for the Hotter'N Hell again. And I seek out sauna the way the Beat poet Allen Ginsberg once confessed that he smoked marijuana, in his epic 1950s poem *America*: "Every chance I get."

But my reasons for doing it have changed. In fact, this journey through the science of humans and heat has shifted my entire approach to health and fitness. I had stumbled into it because I was obsessed with the biological mechanisms governing health and especially aging, the greatest unsolved mystery in biology. (You have my editors to thank for keeping me from diving too far into "the weeds.")

But underlying all of it was my belief that human health *all* boils down to mechanisms—push Lever A, get Outcome B. Go to sauna (or sit in a hot tub) for X minutes at a certain temperature, and you will improve your cardiovascular function and arterial flexibility by a certain amount, leading to a lower risk of heart attack (and stroke), so you will automatically live longer. This stems from a view of the human body as little different from a machine, albeit a somewhat moody and unpredictable machine that can be extremely tricky to repair (not unlike a vintage Land Rover).

This is, essentially, the biohackers' view of the world: All we need to do is tweak this pathway or activate that molecule, and we will reveal the shortcut to better health and better performance. I had gone looking for that shortcut in the realms of sauna and heat exposure, and I had found it. But at a certain point, I also came to the realization that this purely mechanistic, Jiffy Lube kind of view of the world only gets you so far. My sojourn in the hot tube, in Vail, had cracked open a window into another, more important dimension to the story—one that I had overlooked, even though it was right in front of me the whole time.

In Chuck Raison's early studies of hyperthermia for depression, in Switzerland, the investigators noticed something odd: As the depressed study subjects were cooking in the infrared sauna tent, they began talking. Out of nowhere.

"The people in the lab didn't know what to do," Raison says. "They came to us and said, 'The patients are talking! What should we say?'"

The researchers didn't know what to say back, or how to measure this phenomenon, but the story stuck with me. Something about getting hot made these people—these sad, lonely, hopeless people—want to reach out and make contact. I had felt something similar in Colorado: "You were *way* chattier afterward," Ashley Mason told me later.

I have no idea what I might have said (!) in my sauna-drunk state, but I do remember feeling a yearning for connection that went unfulfilled in the clinical environment of the hospital. It had happened in other settings as well—even with the Blister Boys. Although we could only smile and nod at each other, not speaking the same language, I'm pretty sure they weren't in there to boost their heat shock proteins or shave a few points off their blood pressure readings. They were there to sweat, and see friends, and get clean, and talk, and relax, without clothes, in a place that was not work and not home, where they could simply be their true selves.

Sitting in the sauna and sweating, participating in this ancient practice, accomplishes all of those mechanistic good things that we know about—but on a deeper level, it helps us reconnect with what it means to be human. And it points toward a more collaborative, as opposed to competitive, approach to wellness and health.

It's like if you go out for a run. On one level, you are using your muscles and burning off last night's extra dessert, and fortifying your cardiovascular system, and possibly helping extend your healthy lifespan by a little bit. All true. But on a deeper level, that run connects you with your evolutionary history, and the essence of what it means to be human: Our species is defined by our ability to walk (and run) on two legs, to sleuth out prey with our big brains (and kill them with tools), and also to sweat, whether on the trail or in a hot space like a sauna. All of these are human superpowers that helped us conquer the world.

But here's the thing: When our ancestors ran, or sweated, or sat in their version of a sauna, they weren't doing it to become "healthier," they were doing it to survive—and they were also never doing it alone, as I realized when I learned about the woman in the cave.

In 2022, a team of scientists in the Netherlands pulled off an incredible feat: They managed to extract human DNA from a 25,000-year-old piece of jewelry, an elk-tooth pendant that had been found in a cave in Siberia.* When they sequenced the ancient DNA, it identified the object's owner as a woman who had migrated some thousand miles from her homeland to the cave where her pendant had been found.[1]

This represented a huge advance in ancient archaeology. It meant that scientists could now link certain artifacts to specific individuals, and not merely anonymous humans from the distant past. It made ancient people real.

But what really caught my attention was that one of the scientists happened to mention that the DNA had probably come from the woman's sweat: "As a porous material, that tooth was likely soaking in sweat," said lead investigator Marie Soressi, an expert in human origins at the University of Leiden, to *New Scientist* magazine. "It worked like a sponge, pulling in that human DNA and trapping it there for twenty-five thousand years."[2]

That blew my mind. Sweat is elusive, ephemeral stuff. Its job is to evaporate, to disappear. This makes sweat extremely tricky to collect and difficult to study, even in laboratories; that had been the point of

---

* This cave, known as Denisova, had been inhabited by multiple different groups of humans over the millennia—as well as by Neanderthals and a heretofore unknown human species known as "Denisovans," not to mention bears, saber-toothed tigers, and other animals, leaving behind a rich mix of bones and artifacts dating back three hundred thousand years.

the whole "sweat soup" exercise that I had submitted to at KSI. But this pendant, this piece of primitive jewelry, had somehow preserved a sweat sample from twenty-five thousand years ago, a tangible link to our ancient, sweaty past.

Whoever she was, this woman must have sweated an awful lot for her sweat to soak into the elk tooth. She had probably been a nomad, trekking across the Siberian steppes in the heat of summer. Maybe she had helped to hunt the elk herself. Perhaps she had sat in the cave before a roaring fire, warming up until sweat pooled on her chest—like a precursor to a sauna. The pendant, or the necklace, likely never left her body until she died.

I thought about her a lot, as I worked on this book—going through all these different heat experiences, and studying all these different scientific observations about humans in heat.

The thing that really struck me was that, through all of this, she probably never sweated alone. To be by oneself, twenty-five thousand years ago, would have been tantamount to a death sentence. Nearly everything she did would have been done with others in her family, her clan, her tribe—as a team effort, a collaboration. Hunting the elk. Sheltering in the cave. Cooking. Foraging. Traveling. Always together. Always connected.

I thought about the woman in the cave again on a spring Friday morning in New York, when I found myself climbing onto a bench in a hot and spacious sauna—like a very steamy cave—along with about twenty-five other people. While the rest of the city was hustling off to work, we were sitting around in our bathing suits in a place called Othership, in the Flatiron district of Manhattan. And while I am normally quite shy and reserved, I was sharing a deeply personal story with a stranger:

A couple of years after graduating from college, I heard from a school friend who I'll call James. I hadn't seen him in a while, but (as I recall) he had recently suffered some sort of romantic disaster involving a European fashion model. He had retreated to the States and needed a place to stay while he got back on his feet and started a new job. I hosted him for a couple of weeks, and then he moved on. No big deal.

But not long afterward, a nice handwritten thank-you note appeared in the mail, with a surprising postscript: "You are a great friend!" he wrote, underlining it for emphasis.

I remember thinking, *I am?*

I had never thought of myself as an especially "great friend" before. I was probably often an annoying friend, even kind of a pain in the ass. James's note changed my outlook. From then on, I sought to *be a great friend* whenever I could.

The stranger smiled and nodded, hanging on every word. Her name was Sasha, and she looked to be about twenty-five, with pale skin and a smattering of tattoos. We appeared to have little in common apart from being very sweaty and having signed up for a morning "Gratitude" session at Othership. Our guide, a kindly, teddy bear–like tech-world dropout named Mihir, had prompted us to "dig into your memory, to find a story about a time when someone expressed true gratitude to you." Out popped James.

Mihir rang a tiny bell, and it was Sasha's turn to talk about how she had stepped up to manage a girlfriend's wedding, even officiating the ceremony, after the friend's mom had abruptly bailed out. "*You're* a great friend!" I joked, trying to hide the fact that I was choking up a little.

Normally, none of these things would happen on a weekday morning in Manhattan: sweating in public, talking with people who are not our therapists or coworkers, sitting with strangers in a state of near undress. (Crying, maybe.) But strangely, it all felt perfectly natural, as if there were nowhere else that any of us was supposed to be. Which is kind of the point of Othership, one of the more unique sauna spaces anywhere.

Located on the same Manhattan block as the infamous 1980s dance club Limelight, a place where people did other kinds of sharing, back in the day, Othership occupies the center of the Venn diagram between sauna, nightclub, SoulCycle, meditation class, and group therapy. The oblong sauna space, with three tiers of wide benches, does feel a little like a spaceship, by design; the woman at the check-in desk had asked, "Is this your first time *journeying* with us?"

The subdued rose-colored lighting was reminiscent of a club or a cocktail bar, but there was no substance on hand stronger than herbal tea and incense smoke. My fellow journeyers were a mix of ages, genders, and body types, from gym bros and yoga moms to older and more interestingly shaped people. About half were in their twenties and thirties, the rest of us somewhat beyond that, but it didn't seem to matter much.

Othership was born in the backyard of a fortyish Toronto couple named Robbie and Shannon Bent. Robbie had struggled with drugs and alcohol, and gravitated to meditation and "plant medicine," i.e., psychedelics, on his journey to sobriety. On their first date, Shannon had taken him to a bathhouse with a cold plunge. He got hooked—on sauna, and on Shannon. They got married.

When the saunas in Toronto all closed due to Covid, they built their own sauna in their backyard, filled a horse trough with water to serve as a cold plunge, set up some chairs and benches around the yard, and invited their friends and neighbors over, just to reclaim some human contact. Soon dozens of people were coming over every evening to steam in the sauna, bathe in the cold water, and sit around drinking tea and talking or meditating.

Before long, it was a business. The first New York location, this Flatiron space, opened in July 2024, but this was different from other sauna houses in the city. Othership's sauna could hold up to ninety people, and it had a spacious lounge area where guests could just hang out and sip water or tea. And they offered a range of "guided" classes, with meditations, soundbaths, and alcohol-free evening socials that are like sober raves. Most classes involve talking or sharing; some entail therapeutic screaming, which anyone who has lived in New York can relate to. (And, of course, cold-water plunging.) There is even the occasional comedy night. You can also just show up, and move in and out at your own pace.

I had sought out Othership because most of the saunas that I had visited outside Finland had been hyper-focused on the purely physical health benefits of sauna, as a "wellness protocol." Which is great, but I wanted something more fulfilling. As author Mikkel Aaland observes, true sauna has three important dimensions: it is physical, of course, but also social, and even spiritual. Othership emphasizes all three, but especially leans into the social side, positioning itself as an antidote to our modern epidemic of loneliness. Something about the heat, Robbie had noticed—perhaps the heat *and* the cold—seems to unlock people, freeing their emotions, their tongues, lowering their guard. Even, maybe especially, in New York.

"If you come to Othership and sit there with a person on your left and a person on your right, you're probably going to talk to them," he says.

Something about the heat seems to bring us out of our shells. The fact that this seems novel, let alone the basis for a whole category of businesses, says a lot about where modern life has gone off the rails.

Back when the sauna or *hammam* or *banya* or sweat lodge did the job of bathtub and shower, as well as whatever ritual and social purposes it was assigned—one would never sweat alone. Even if you did fire up the sauna on your own, you would not be alone for long, as neighbors and friends smelled the smoke and drifted over for a steam. The woman in the cave never sweated alone either. Nor do the Blister Boys, even if some of them might live alone.

Now, too many of us spend way too much time by ourselves, by choice or not. Could heat therapy help cure our loneliness?

While I still sometimes enjoy a meditative solo sauna session, especially when immersed in a creative project, I've found that sweating alone is always so much more difficult and painful for me—not to mention boring.

With others, it always seemed easier to take the heat. Riding with the Texas Turtles (and about nine thousand other like-minded nutjobs) was the only thing that got me through the Hotter'N Hell. Sitting in a hot sauna with the Blister Boys or the Business Slalom revelers was no problem; opposite as they were in some ways, both experiences were soul-satisfying on some level. Hot yoga was the same; I'd never want to do *that* alone. And just about the only way I can handle cold plunging is with company.

In the age before air-conditioning, even hot spells used to bring people together, as the travel writer Jan Morris wrote in *Manhattan '45*, her evocative portrait of the city at the end of World War II: "When a heat wave left the whole city gasping and sweating, a powerful fellowship blunted the edge of the common misery, bridging the most insuperable linguistic barriers, or the most unclimbable social barricades, if only with a wink or a grimace."[3]

I had experienced something similar, at the Hotter'N Hell and in saunas around the world—from smoky trailers parked by the side of a Midwestern lake, to the simple benches of Rajaportti, to the nightlife saunas of New York, to a $500 million bathing palace in Eastern Europe called Therme Bucharest that felt like a cross between the Roman Baths of Caracalla and Disney World. At Therme, I experienced the apotheosis of sauna culture in an art form called *aufguss*—brief dramatic performances conducted in 180 degree F sauna spaces. In one, I sat with fifty other people as a lean, bald Norwegian singer embraced us all in a tragic aria ("E lucevan le stelle") from the opera *Tosca*, leaving not a dry eye in the house.

In each of these places, in different ways, I had felt the power of heat to heal and connect us, to one another and ourselves.

But I found the most powerful iteration of heat bathing in perhaps the least likely spot: A suburban backyard in Draper, Utah.

Nestled at the base of the Wasatch Mountains, just south of Salt Lake City, Draper is a picture-perfect suburb comprised of very similar, beigey homes. On a Wednesday afternoon, I found my way to a house that was not like the others, because it happened to have a homemade sweat lodge in the backyard, a low domed structure about four feet tall.

The home belonged to a very intense massage therapist named Eric, a man with a fiery red beard, who was so passionate about heat bathing that he had built this backyard sweat lodge before finishing the railings on his deck. It had become a gathering place for an eclectic group of military veterans, spiritual seekers, yoga instructors, and other heat addicts. I was invited by an ex-Army Ranger of my acquaintance named Jeff Kirkham, a veteran of Afghanistan and Iraq who lives next door.

One of Jeff's Army friends also showed up, pointed to a hole that city workmen had dug in the street out front, and joked, "That an IED?"

One by one, we crawled in through a low doorway covered by a blanket. It was cramped inside, too low to stand up. We took our places around a simple camp stove, fueled by Duraflame logs (which burn hotter and cleaner than wood), with a saucepan full of water on top to provide steam. The pot also held a crystal the size of a baby's head.

Though a seasoned heat-seeker by now, I had read enough "Man Dies in Sweat Lodge" articles online to feel apprehensive. This one looked a little sketchy compared to the posh bathing emporia that I had been frequenting. The thermometer on the wall only went up to 130 degrees F, and the needle was pinned well past that. Not satisfied, Eric threw another half a Duraflame into the stove. It started to feel hot and crowded, with seven of us in the tiny space. As my claustrophobia began to rise along with the temperature, I focused on the sliver of daylight and grass through the tiny gap in the blanket covering the door.

"Don't pass out," Jeff warned, "or we'll have to drag your body out and throw you in the cold plunge."

Led by Eric, we did some ritual-ish stuff that I don't precisely recall; there was "gratitude," and sharing, and some scraping, using dull metal instruments to massage various body parts. After three 20-minute rounds, we all crawled out for one last plunge, in a Tractor Supply tub filled by a garden hose.

It was twilight, and as I approached the tub, I saw that it was occupied. And vibrating. Jeff was in there, making a guttural rumbling sound, with his eyes closed and his hands clasped in front of his face. Without hesitating, I climbed in beside him.

After that, we all lay down on mats on the grass for breathing exercises. Which, if I'm being honest, I've always kind of thought was silly. But I went with the flow. One of the yoga teachers handed out wireless headphones, so we could follow her cues: two sharp inhales (*shhh-shhh!*) followed by a longer exhale (*haaaaahh*). Then again. And again. Quickly. My brain filled up with oxygen. I started to feel lightheaded.

Sshh-sshh-haaaaahh. Sshh-sshh-haaaaahh. Sshh-sshh-haaaaahh.

I lay on my back staring at the sky. The breathwork cues gave way to music, a string orchestra playing swelling, inspiring music that I eventually recognize as Madonna's 1980s hit, "Material Girl"—which, normally, my body would reject like bad sushi. But in my sauna-drunk state, I hummed along, feeling the music lift me into the air, through the trees, with the birds in the sky.

Everything just seemed to *sloooowwww dowwwwnnn*.

"Does anyone else feel high?" someone asked. Yep.

I was, finally, sauna drunk. In *Totonou*, the mythical Japanese post-sauna trance. It was real. The music soared. The trees swayed gently. Birds flitted across my field of view, soaring and swooping to freedom, like my thoughts. *Sshh-sshh-haaaaahh. Sshh-sshh-haaaaahh.*

It was almost dark by the time I peeled myself off the mat. I said goodbye to Jeff and Eric, and everyone else, then staggered to my car to drive home, my head still floating around with the birds. As I pulled onto the main road, I was apparently going too slow for some of the other drivers, who honked and veered around me in anger.

Blissed out, and grooving on *totonou*, I flipped them a languid middle finger and drove on home.

ACKNOWLEDGMENTS

This book began as an assignment for *Men's Health* magazine to investigate the mysteries of human sweat and heat adaptation—two topics that, if I'm being honest, I might have overlooked. Thanks to my editors Ben Paynter and Ben Court for sending me on a long and life-changing journey.

It quickly became clear that this subject was far too big and fascinating for a brief magazine article, and my enterprising agent, Daniel Greenberg, seized the moment to make this book happen. There was no editor I'd rather work with than Diana Baroni, who immediately said yes to a book idea that boiled down to a single word—*heat*. Diana helped me shape a broad topic into a coherent story, patiently pointing out the bits that weren't working and letting me fix them.

A book like this could never be written without the cooperation of many scientists and experts, who generously shared their expertise and their precious time. At the Korey Stringer Institute of the University of Connecticut, Douglas Casa, Robert Huggins, and Rebecca Stearns welcomed me into their lab and shared their knowledge and passion for heat training and heat safety. Huggy Bear (Huggins) expertly guided me through the journey to heat adaptation and patiently explained the science behind it.

ACKNOWLEDGMENTS

Christopher Minson of the University of Oregon first turned me on to the magic of sweat and heat adaptation, the topic to which he's devoted his career. Chuck Raison (Wisconsin) and Ashley Mason (UCSF) opened my eyes to the possible mental-health applications of heat therapy, and helped me get through my stint as a test subject.

Thanks also to the other scientists and experts who offered insights and expertise, including: Anthony Bain, Simmie Foster, Chris Lowry, Kathleen Fisher, Lindsay Baker, Mike Tipton, Louise Burke, Paul Laursen, Patrick Brown, Meridith Cass, Andrew Best, Yana Kamberov, Daniel Lieberman, Evan Johnson, Andreas Flouris, Bob Roemer, Ross Tucker, and Tim Noakes, among others. Among athletes, thanks especially to CJ Albertson, Sylvia Bedford, Patrick Frigge, Meridith Cass, and Olav Aleksander Bu.

Rhonda Patrick has done more than anyone else to interpret and popularize the once-obscure scientific literature around sauna and heat adaptation for performance and health. We never spoke, but her work has created much public awareness around this topic, and everyone interested in heat therapy, heat adaptation, and the health implications of sauna owes her a debt.

David Martin and Sean Langan not only handheld me through heat testing at KSI, but later provided invaluable research and insights into my drafts at various points in the process. Sean pulled an especially heavy lift with the many, many notes and sources, while David acted as part coach and part babysitter while I trained for the Hotter'N Hell Hundred.

In the wonderful world of sauna bathing enthusiasts, I have been lucky to share the bench with Eero Kilpi, Glenn Auerbach, Justin Juntunen, Annette Scott, Robert Hammond, and the sauna godfather, Mikkel Aaland, among many others. In Finland, Risto Elomaa proved to be a patient and knowledgeable sauna shepherd, as did the incomparable and passionate Alex Lembke at Rajaportti. Thanks also to Anu Puustinen, Chef Kim Mustonen, the Business Slalom gang, and the Blister Boys. Thanks to Visit Finland, Visit Tampere, the Hotel Maria, and the Panorama Landscape Hotel for help and guidance.

I owe beers, or more, to my beta readers for getting through my shitty first drafts with thoughtfulness—and above all, tact. Your insights proved vital in refining and improving the narrative. Thanks especially to Peter O'Toole, Brian Durbin, Stephen Dark, David Martin, Ted Nordhaus, Tyler Schmidt, and especially Alex and Susan Heard, whose late-inning polish helped me sleep better at night.

Rodrigo Corral and his team, the brilliant designers behind *Outlive*, delivered yet another iconic cover. I am truly honored to be in their pantheon of authors. Eo Trueblood created a beautiful rendering of the humble eccrine sweat gland. Alex Lembke graciously allowed you to see the evocative sauna photos on pages 168 and 169.

Thanks also to the team at Harper Wave, including Odette Fleming, Megan Wilson, Emma Effinger, Rachel Weinick, Melissa Lotfy, Owen Corrigan, Marleen Reimer, and Emma Gilheany.

Sarah Gifford, a talented book designer who also happens to be my sister, helped me navigate and understand the different visual options. Andrew Murphy kept me laughing, humble, and motivated; also, he and his friend Raphael Andreae braved the saunas of Therme Erding (Germany) on my behalf.

Thanks most of all to Martha McGraw for always being there for me, for creating space for me to work and travel, for putting up with the ups and downs of the writing life, for being incredibly brave and strong, and for being the perfect traveling companion to Finland and elsewhere. (Let's go back! In summer!)

Bill Gifford, Salt Lake City, October 2025

NOTES

INTRODUCTION

1. Rollins H. 2011. "The Rollins Column: August Ode." *LA Weekly*. https://www.laweekly.com/henry-rollins-the-column-august-ode/.
2. IPCC. 2023. *Climate Change 2023: Synthesis Report. Contribution of Working Groups I, II and III to the Sixth Assessment Report of the Intergovernmental Panel on Climate Change* [Core Writing Team, H. Lee and J. Romero (eds.)].
3. Tao L, Su Y, Chen X, Tian F. 2024. "Processes, spatial patterns, and impacts of the 1743 extreme-heat event in northern China: from the perspective of historical documents." *Climate of the Past* 20 (11): 2455–2471. https://doi.org/10.5194/cp-20-2455-2024. https://cp.copernicus.org/articles/20/2455/2024/.
4. U.S. Department of Health and Human Services. 2022. "Heat.gov." Accessed July 8. https://www.heat.gov.
5. Aaland M. 1978. *Sweat: the illustrated history and description of the Finnish sauna, Russian bania, Islamic hammam, Japanese mushi-buro, Mexican temescal, and American Indian & Eskimo sweat lodge*. Capra Press.

1. BURNING MAN

1. Cottle RM, Fisher KG, Leach OK, Wolf ST, Kenney WL. 2024. "Critical environmental core temperature limits and heart rate thresholds across the adult age span (PSU HEAT Project)." *Journal of Applied Physiology* 137 (1): 145–153. https://doi.org/10.1152/japplphysiol.00117.2024. https://journals.physiology.org/doi/10.1152/japplphysiol.00117.2024.

2. Schneider SM. 2016. "Heat acclimation: Gold mines and genes." *Temperature (Austin)* 3 (4): 527–538. https://doi.org/10.1080/23328940.2016.1240749. http://www.ncbi.nlm.nih.gov/pubmed/28090556.
3. Kupperman KO. 1984. "Fear of Hot Climates in the Anglo-American Colonial Experience." *The William and Mary Quarterly* 41 (2): 213. https://doi.org/10.2307/1919050. https://www.jstor.org/stable/1919050?origin=crossref.
4. Levick JJ. 1859. "Remarks on sunstroke." *American Journal of Medicine* 37 (43).

2. SWEAT

1. Blagden, C. "Experiments and Observations in an Heated Room By Charles Blagden, M. D. F. R. S." *Philosophical Transactions* (1683–1775) 65 (1775): 111–23. http://www.jstor.org/stable/106183.
2. Kupperman KO. 1984. "Fear of Hot Climates in the Anglo-American Colonial Experience." *The William and Mary Quarterly* 41 (2): 213. https://doi.org/10.2307/1919050. https://www.jstor.org/stable/1919050.
3. Sambon LW. 1898. "Acclimatization of Europeans in Tropical Lands." *The Geographical Journal* 12 (6): 589–599.
4. Blagden, C. "Experiments and Observations in an Heated Room."
5. Maron MB, Wagner JA, Horvath SM. 1977. "Thermoregulatory responses during competitive marathon running." *Journal of Applied Physiology: Respiratory, Environmental and Exercise Physiology* 42 (6): 909–914. https://doi.org/10.1152/jappl.1977.42.6.909. https://www.ncbi.nlm.nih.gov/pubmed/881391; also see Racinais S, Moussay S, Nichols D, et al. 2019. "Core temperature up to 41.5ºC during the UCI Road Cycling World Championships in the heat." *British Journal of Sports Medicine* 53 (7): 426–429. https://doi.org/10.1136/bjsports-2018-099881.
6. Maron MB, et al. "Thermoregulatory responses during competitive marathon running." Maron MB. 2014. "The University of California Institute of Environmental Stress marathon field studies." *Advances in Physiology Education* 38 (1): 3–11. https://doi.org/10.1152/advan.00118.2013.
7. Hardy JD, Du Bois EF, Soderstrom GF. 1938. Basal metabolism, radiation, convection and vaporization at temperatures of 22 to 35 C.: Six figures." *The Journal of Nutrition* 15.5 (1938): 477–497. https://doi.org/10.1093/jn/15.5.477.
8. Lu C, Fuchs E. 2014. "Sweat gland progenitors in development, homeostasis, and wound repair." *Cold Spring Harbor Perspectives in Medicine* 4 (2): a015222. https://doi.org/10.1101/cshperspect.a015222.
9. Romanovsky AA. 2018. "The thermoregulation system and how it works." *Handbook of Clinical Neurology* 156: 3–43. https://doi.org/10.1016/B978-0-444-63912-7.00001-1.
10. Lieberman DE. 2014. "Human Locomotion and Heat Loss: An Evolutionary Perspective." In *Comprehensive Physiology*, edited by Terjung R, 99–117. Wiley.
11. Best A, Lieberman DE, Kamilar JM. 2019. "Diversity and evolution of human eccrine sweat gland density." *Journal of Thermal Biology* 84: 331–338. https://doi.org/10.1016/j.jtherbio.2019.07.024.

12. Whitford WG. 1976. "Sweating responses in the chimpanzee (Pan troglodytes)." *Comparative Biochemistry and Physiology Part A: Physiology* 53 (4): 333–336. https://doi.org/10.1016/S0300-9629(76)80151-X. https://linkinghub.elsevier.com/retrieve/pii/S030096297680151X; Kamberov YG, Guhan SM, DeMarchis A, et al. 2018. "Comparative evidence for the independent evolution of hair and sweat gland traits in primates." *Journal of Human Evolution* 125: 99–105. https://doi.org/10.1016/j.jhevol.2018.10.008.
13. Kamberov YG, Karlsson EK, Kamberova GL, et al. 2015. "A genetic basis of variation in eccrine sweat gland and hair follicle density." *Proceedings of the National Academy of Sciences of the United States of America* 112 (32): 9932–9937. https://doi.org/10.1073/pnas.1511680112.
14. Aldea D, Kamberov YG. 2022. "En1 sweat we trust: How the evolution of an Engrailed 1 enhancer made humans the sweatiest ape." *Temperature (Austin)* 9 (4): 303–305. https://doi.org/10.1080/23328940.2021.2019548.
15. Andersson R. 2015. "Promoter or enhancer, what's the difference? Deconstruction of established distinctions and presentation of a unifying model." *BioEssays* 37 (3): 314–323. https://doi.org/10.1002/bies.201400162.
16. Bramble DM, Lieberman DE. 2004. "Endurance running and the evolution of Homo." *Nature* 432 (7015): 345–352. https://doi.org/10.1038/nature03052.
17. Kricho, Kade. "China's Stone Age Skiers and History's Harsh Lessons." *The New York Times*, April 19, 2017. https://www.nytimes.com/2017/04/19/sports/skiing/skiing-china-cave-paintings.html.

3. GOOD HEAT

1. Zech M, Bosel S, Tuthorn M, et al. 2015. "Sauna, sweat and science - quantifying the proportion of condensation water versus sweat using a stable water isotope ((2)H/(1)H and (18)O/(16)O) tracer experiment." *Isotopes in Environmental and Health Studies* 51 (3): 439–447. https://doi.org/10.1080/10256016.2015.1057136.
2. Leppäluoto J. 1988. "Human thermoregulation in sauna." *Annals of Clinical Research* 20 (4): 240–243. http://www.ncbi.nlm.nih.gov/pubmed/3218894.
3. University of Eastern Finland. "Kuopio Ischaemic Heart Disease Risk Factor (KIHD) Study, Nutrition." University of Eastern Finland. Accessed July 10. https://uefconnect.uef.fi/en/kuopio-ischaemic-heart-disease-risk-factor-kihd-study-nutrition/#publications.
4. Foster H. 1976. "Heart-attacks and the Sauna." *The Lancet* 308 (7980): 313. https://doi.org/10.1016/S0140-6736(76)90765-0.
5. Patrick R. "Dr. Jari Laukkanen on Sauna Use for the Prevention of Cardiovascular & Alzheimer's Disease." Accessed May 20, 2023. https://www.foundmyfitness.com/episodes/jari-laukkanen.
6. Laukkanen T, Khan H, Zaccardi F, Laukkanen JA. 2015. "Association Between Sauna Bathing and Fatal Cardiovascular and All-Cause Mortality Events." *JAMA Internal Medicine* 175 (4): 542. https://doi.org/10.1001/jamainternmed.2014.8187.

7. Kunutsor, S. K., Khan, H., Zaccardi, F., Laukkanen, T., Willeit, P., Laukkanen, J. A. 2018. "Sauna bathing reduces the risk of stroke in Finnish men and women: A prospective cohort study." *Neurology*, 90(22), e1937–e1944. https://doi.org/10.1212/WNL.0000000000005606.
8. Laukkanen T, Kunutsor S, Kauhanen J, Laukkanen JA. 2017. "Sauna bathing is inversely associated with dementia and Alzheimer's disease in middle-aged Finnish men." *Age and Ageing* 46 (2): 245–249. https://doi.org/10.1093/ageing/afw212.
9. Laukkanen T, Laukkanen JA, Kunutsor SK. 2018. "Sauna Bathing and Risk of Psychotic Disorders: A Prospective Cohort Study." *Medical Principles and Practice* 27 (6): 562–569. https://doi.org/10.1159/000493392.
10. Kunutsor SK, Khan H, Laukkanen T, Laukkanen JA. 2018. "Joint associations of sauna bathing and cardiorespiratory fitness on cardiovascular and all-cause mortality risk: a long-term prospective cohort study." *Annals of Medicine* 50 (2): 139–146. https://doi.org/10.1080/07853890.2017.1387927.
11. Tuomilehto, J., Sarti, C., Narva, E. V., Salmi, K., Sivenius, J., Kaarsalo, E., Salomaa, V., Torppa, J. 1992. The FINMONICA Stroke Register. Community-based stroke registration and analysis of stroke incidence in Finland, 1983–1985. *American Journal of Epidemiology*, 135(11), 1259–1270. https://doi.org/10.1093/oxfordjournals.aje.a116232.
12. Ganga, A., Jayaraman, M. V., E Santos Fontánez, S., Moldovan, K., Torabi, R., & Wolman, D. N. (2024). Population analysis of ischemic stroke burden and risk factors in the United States in the pre- and post-mechanical thrombectomy eras. *Journal of Stroke and Cerebrovascular Diseases: The Official Journal of National Stroke Association*, 33(8), 107768. https://doi.org/10.1016/j.jstrokecerebrovasdis.2024.107768.
13. Hill AB. 1965. "The Environment and Disease: Association or Causation?" *Proceedings of the Royal Society of Medicine* 58 (5): 295–300. https://doi.org/10.1177/003591576505800503.
14. Knekt P, Järvinen R, Rissanen H, Heliövaara M, Aromaa A. 2020. "Does sauna bathing protect against dementia?" *Preventive Medicine Reports* 20: 101221. https://doi.org/10.1016/j.pmedr.2020.101221.
15. Laukkanen JA, Laukkanen T, Kunutsor SK. 2018. "Cardiovascular and Other Health Benefits of Sauna Bathing: A Review of the Evidence." *Mayo Clinic Proceedings* 93 (8): 1111–1121. https://doi.org/10.1016/j.mayocp.2018.04.008; Lee E, Kolunsarka I, Kostensalo J, et al. 2022. "Effects of regular sauna bathing in conjunction with exercise on cardiovascular function: a multi-arm, randomized controlled trial." *American Journal of Physiology: Regulatory, Integrative and Comparative Physiology* 323 (3): R289–R299. https://doi.org/10.1152/ajpregu.00076.2022.
16. Patrick RP, Johnson TL. 2021. "Sauna use as a lifestyle practice to extend healthspan." *Experimental Gerontology* 154: 111509. https://doi.org/10.1016/j.exger.2021.111509.
17. Leppäluoto J, Huttunen P, Hirvonen J, Väänänen A, Tuominen M, Vuori J. 1986. "Endocrine effects of repeated sauna bathing." *Acta Physiologica Scandinavica* 128 (3): 467–470. https://doi.org/10.1111/j.1748-1716.1986.tb08000.x.

18. Jezova D, Vigas M, Tatar P, Jurcovicova J, Palat M. 1985. "Rise in plasma beta-endorphin and ACTH in response to hyperthermia in sauna." *Hormone and Metabolic Research* 17 (12): 693–694. https://doi.org/10.1055/s-2007-1013648.
19. Ritossa F. 1962. "A New Puffing Pattern Induced by Temperature Shock and DNP in Drosophila." *Experientia*: 571–573. https://doi.org/https://doi.org/10.1007/BF02172188; Feder ME, Hofmann GE. 1999. "Heat-shock proteins, molecular chaperones, and the stress response: evolutionary and ecological physiology." *Annual Review of Physiology* 61: 243–82. https://doi.org/10.1146/annurev.physiol.61.1.243.
20. Ely BR, Clayton ZS, McCurdy CE, et al. 2019. "Heat therapy improves glucose tolerance and adipose tissue insulin signaling in polycystic ovary syndrome." *American Journal of Physiology, Endocrinology and Metabolism* 317 (1): E172-E182. https://doi.org/10.1152/ajpendo.00549.2018.
21. Johnson CN, Jensen RS, Von Schulze AT, Geiger PC. 2022. "Heat Therapy Can Improve Hepatic Mitochondrial Function and Glucose Control." *Exercise and Sport Sciences Reviews* 50 (3): 162–170. https://doi.org/10.1249/JES.0000000000000296.
22. Olsen, A., Vantipalli, M. C., & Lithgow, G. J. (2006). "Lifespan extension of Caenorhabditis elegans following repeated mild hormetic heat treatments." *Biogerontology*, 7(4), 221–230. https://doi.org/10.1007/s10522-006-9018-x; P. Sarup, P. Sørensen, V. Loeschcke, 2014. "The long-term effects of a life-prolonging heat treatment on the Drosophila melanogaster transcriptome suggest that heat shock proteins extend lifespan." *Experimental Gerontology*, Volume 50, 2014, 34-39. https://doi.org/10.1016/j.exger.2013.11.017.
23. Tei C, Shinsato T, Kihara T, Miyata M. 2006. "Successful thermal therapy for end-stage peripheral artery disease." *Journal of Cardiology* 47 (4): 163–4. https://www.ncbi.nlm.nih.gov/pubmed/16637249.
24. Ukai T, Iso H, Yamagishi K, et al. 2020. "Habitual tub bathing and risks of incident coronary heart disease and stroke." *Heart* 106 (10): 732–737. https://doi.org/10.1136/heartjnl-2019-315752.
25. Brunt VE, Howard MJ, Francisco MA, Ely BR, Minson CT. 2016. "Passive heat therapy improves endothelial function, arterial stiffness and blood pressure in sedentary humans." *The Journal of Physiology* 594 (18): 5329–5342. https://doi.org/10.1113/JP272453.

4. TOO COOL

1. Data on AC prevalence in New York City from the NYC Department of Health Environment & Health Data Portal: https://a816-dohbesp.nyc.gov/IndicatorPublic/data-stories/heat/.
2. Miller A. 1998. "Before Air-Conditioning." *The New Yorker*.
3. Stouhi D. 2021. "What Is a Traditional Windcatcher?" *Arch Daily* blog. https://www.archdaily.com/971216/what-is-a-traditional-windcatcher.
4. Frank Lloyd Wright. 1954. *The Natural House*. Horizon Press. p. 178.
5. Ritchie H. 2024. "Air conditioning causes around 3% of greenhouse gas emissions. How will this change in the future?" *Our World in Data*. Accessed July 22,

2025. https://ourworldindata.org/air-conditioning-causes-around-greenhouse-gas-emissions-will-change-future.
6. Matthew Dalton. "The New Hot Topic in European Politics is Air Conditioning." *The Wall Street Journal.* July 22, 2025. https://www.wsj.com/world/europe/europe-air-condition-heat-waves-politics-24aceab4.
7. Dixon PG, Mote TL. 2003. "Patterns and Causes of Atlanta's Urban Heat Island–Initiated Precipitation." *Journal of Applied Meteorology and Climatology* 42 (9): 1273–1294. https://doi.org/10.1175/1520-0450(2003)042<1273:PACOAU>2.0.CO;2. See also Bornstein, R. and Q. Lin. 2000. "Urban heat islands and summertime convective thunderstorms in Atlanta: Three case studies." *Atmos. Environ.* 34:507–516.
8. Spelman College. 2022. "UrbanHeatATL." Accessed July 11, 2024. https://urbanheatatl.org.
9. Matheus Gouvea de Andrade. 2023. "How Medellin is beating the heat with green corridors." BBC. Accessed July 17. https://www.bbc.com/future/article/20230922-how-medellin-is-beating-the-heat-with-green-corridors; Alcaldia de Medellin Secretaria de Medio Ambiente. "Corredores y muros verdes." Accessed July 22. https://www.medellin.gov.co/es/secretaria-medio-ambiente/medellin-biodiversa/corredores-y-muros-verdes/; Lisa W. Foderaro. 2015. "Bronx planting caps off a drive to add a million trees." *The New York Times.* October 20, 2015. https://www.nytimes.com/2015/10/21/nyregion/new-york-city-prepares-to-plant-one-millionth-tree-fulfilling-a-promise.html?ref=nyregion.
10. Smart R. 2022. "Ditch the tie and reduce the AC—Japan's Cool Biz gets summer hell just about right." *Quartz.* Accessed July 22. https://qz.com/465327/ditch-the-tie-and-reduce-the-ac-japans-cool-biz-gets-summer-hell-just-about-right.
11. Downs S. 2022. "A Survey of Modern Life: Outdoor Time." Medium. https://medium.com/building-h/a-survey-of-modern-life-outdoor-time-3a99d9fa3acb. DJ Case and Associates. 2025. "The Nature of Americans." https://natureofamericans.org.
12. Everts S. 2021. "The Joy of Sweat." W. W. Norton & Company.

5. ADAPTATION

1. Ferriss T, Patrick R. Are Saunas the Next Big Performance-Enhancing "Drug"? *Tim Ferriss* blog. https://tim.blog/2014/04/10/saunas-hyperthermic-conditioning-2/.
2. Gay J. 2024. "Chicken Coop Heat Lamps and 2:09 on a Treadmill: Meet the 'Mad Scientist' of Marathoning." *The Wall Street Journal.* https://www.wsj.com/sports/cj-albertson-new-york-city-marathon-f8d3e0f9.
3. Mantzios K, Ioannou LG, Panagiotaki Z, et al. 2022. "Effects of Weather Parameters on Endurance Running Performance: Discipline-specific Analysis of 1258 Races." *Med Sci Sports Exerc* 54 (1): 153–161. https://doi.org/10.1249/MSS.0000000000002769.
4. Hosokawa Y, Stearns RL, Casa DJ. 2019. "Is Heat Intolerance State or Trait?" *Sports Med* 49 (3): 365–370. https://doi.org/10.1007/s40279-019-01067-z.
5. Tucker R, Rauch L, Harley YX, Noakes TD. 2004. "Impaired exercise performance in the heat is associated with an anticipatory reduction in skeletal muscle recruit-

ment." *Pflugers Archiv* 448 (4): 422–430. https://doi.org/10.1007/s00424-0
04-1267-4.
6. Fotheringham W. 2007. *Put Me Back on My Bike: In Search of Tom Simpson*. London: Yellow Jersey Press.
7. Ely MR, Cheuvront SN, Roberts WO, Montain SJ. 2007. "Impact of Weather on Marathon-Running Performance." *Medicine & Science in Sports & Exercise* 39 (3): 487–493. https://doi.org/10.1249/mss.0b013e31802d3aba.
8. Hosier G. How Much Does Heat Slow Down Your Race Pace? *Outside Online* blog. 2019. https://www.outsideonline.com/health/running/racing/race-strategy/how-much-does-heat-slow-your-race-pace/; El Helou N, Tafflet M, Berthelot G, et al. 2012. "Impact of Environmental Parameters on Marathon Running Performance." *PLoS ONE* 7 (5): e37407. https://doi.org/10.1371/journal.pone.00 37407; Hutchinson A. 2021. *Endure: Mind, Body, and the Curiously Elastic Limits of Human Performance*. Revised and updated ed. New York: Custom House.
9. Carr AJ, Vallance BS, Rothwell J, Rea AE, Burke LM, Guy JH. 2022. "Competing in Hot Conditions at the Tokyo Olympic Games: Preparation Strategies Used by Australian Race Walkers." *Front Physiol* 13: 836858. https://doi.org/10.3389/fphys.2022.836858.
10. Munsters C, Siegers E, Sloet van Oldruitenborgh-Oosterbaan M. 2024. "Effect of a 14-Day Period of Heat Acclimation on Horses Using Heated Indoor Arenas in Preparation for Tokyo Olympic Games." *Animals (Basel)* 14 (4): 546. https://doi.org/10.3390/ani14040546.
11. Gibson OR, Tuttle JA, Watt PW, Maxwell NS, Taylor L. 2016. "Hsp72 and Hsp90alpha mRNA transcription is characterised by large, sustained changes in core temperature during heat acclimation." *Cell Stress and Chaperones* 21 (6): 1021–1035. https://doi.org/10.1007/s12192-016-0726-0; Fox RH, Goldsmith R, Kidd DJ, Lewis HE. 1963. "Acclimatization to heat in man by controlled elevation of body temperature." *J Physiol* 166 (3): 530–547. https://doi.org/10.1113/jphysiol.1963.sp007121.
12. Wyndham CH. 1967. "Effect of acclimatization on the sweat rate-rectal temperature relationship." *J Appl Physiol* 22 (1): 27–30. https://doi.org/10.1152/jappl.1967.22.1.27.
13. Buono, M. J., Kolding, M., Leslie, E., Moreno, D., Norwood, S., Ordille, A., & Weller, R. (2018). Heat acclimation causes a linear decrease in sweat sodium ion concentration. *Journal of Thermal Biology*, 71, 237–240. https://doi.org/10.1016/j.jtherbio.2017.12.001.
14. Wyndham CH. 1977. "Heat stroke and hyperthermia in marathon runners." *Ann N Y Acad Sci* 301: 128–38. https://doi.org/10.1111/j.1749-6632.1977.tb38192.x; Maron MB, et al., 1977. "Thermoregulatory responses during competitive marathon running." *J Appl Physiol Respir Environ Exerc Physiol* 42 (6): 909–14. https://doi.org/10.1152/jappl.1977.42.6.909.
15. Kenefick RW, Cheuvront SN, Palombo LJ, Ely BR, Sawka MN. 2010. "Skin temperature modifies the impact of hypohydration on aerobic performance." *J Appl Physiol* (1985) 109 (1): 79–86. https://doi.org/10.1152/japplphysiol.00135.2010. https://www.ncbi.nlm.nih.gov/pubmed/20378704.

16. Scoon GS, Hopkins WG, Mayhew S, Cotter JD. 2007. Effect of post-exercise sauna bathing on the endurance performance of competitive male runners. *Journal of Science and Medicine in Sport*, 10(4), 259–262. https://doi.org/10.1016/j.jsams.2006.06.009.
17. Lundby C, Hamarsland H, Hansen J, et al. 2023. "Hematological, skeletal muscle fiber, and exercise performance adaptations to heat training in elite female and male cyclists." *J Appl Physiol* (1985) 135 (1): 217–226. https://doi.org/10.1152/japplphysiol.00115.2023. https://www.ncbi.nlm.nih.gov/pubmed/37262101.
18. Ronnestad BR, Hamarsland H, Hansen J, et al. 2021. "Five weeks of heat training increases haemoglobin mass in elite cyclists." *Exp Physiol* 106 (1): 316–327. https://doi.org/10.1113/EP088544. https://www.ncbi.nlm.nih.gov/pubmed/32436633.
19. Baranauskas MN, Constantini K, Paris HL, Wiggins CC, Schlader ZJ, Chapman RF. 2021. "Heat Versus Altitude Training for Endurance Performance at Sea Level." *Exerc Sport Sci Rev* 49 (1): 50–58. https://doi.org/10.1249/JES.0000000000000238. https://www.ncbi.nlm.nih.gov/pubmed/33044330. https://journals.lww.com/acsm-essr/fulltext/2021/01000/heat_versus_altitude_training_for_endurance.7.aspx.
20. Caldwell AR, Oki K, Ward SM, et al. 2021. "Impact of successive exertional heat injuries on thermoregulatory and systemic inflammatory responses in mice." *J Appl Physiol* (1985) 131 (5): 1469–1485. https://doi.org/10.1152/japplphysiol.00160.2021. https://www.ncbi.nlm.nih.gov/pubmed/34528459.
21. Ebisuda, Y., Mukai, K., Takahashi, Y., Yoshida, T., Matsuhashi, T., Kawano, A., Miyata, H., Kuwahara, M., & Ohmura, H. (2024). Heat acclimation improves exercise performance in hot conditions and increases heat shock protein 70 and 90 of skeletal muscles in Thoroughbred horses. *Physiological Reports*, 12, e16083. https://doi.org/10.14814/phy2.16083.
22. Gomez Isaza DF, Rodgers EM. 2022. "Exercise training does not affect heat tolerance in Chinook salmon (Oncorhynchus tshawytscha)." *Comp Biochem Physiol A Mol Integr Physiol* 270: 111229. https://doi.org/10.1016/j.cbpa.2022.111229.
23. Shein, Na'ama A et al. "Heat acclimation: a unique model of physiologically mediated global preconditioning against traumatic brain injury." *Progress in Brain Research* vol. 161 (2007): 353–63. doi:10.1016/S0079-6123(06)61025-X.
24. Brunt VE, Howard MJ, Francisco MA, Ely BR, Minson CT. 2016. "Passive heat therapy improves endothelial function, arterial stiffness and blood pressure in sedentary humans." *The Journal of Physiology* 594 (18): 5329–5342. https://doi.org/10.1113/JP272453.
25. Brunt VE, Howard MJ, Francisco MA, Ely BR, Minson CT. 2016. "Passive heat therapy improves endothelial function, arterial stiffness and blood pressure in sedentary humans." *The Journal of Physiology* 594 (18): 5329–5342. https://doi.org/10.1113/JP272453.
26. Pollak A, Merin G, Horowitz M, Shochina M, Gilon D, Hasin Y. 2017. "Heat Acclimatization Protects the Left Ventricle from Increased Diastolic Chamber Stiffness Immediately after Coronary Artery Bypass Surgery: A Lesson from 30 Years of Studies on Heat Acclimation Mediated Cross Tolerance." *Front Physiol* 8: 1022. https://doi.org/10.3389/fphys.2017.01022; Horowitz M, Hasin Y. 2023. "Vascular compliance and left ventricular compliance cross talk: Implications for using

long-term heat acclimation in cardiac care." *Front Physiol* 14: 1074391. https://doi.org/10.3389/fphys.2023.1074391.

6. DRINKING PROBLEM

1. Cheuvront SN, Kenefick RW. 2016. "Am I Drinking Enough? Yes, No, and Maybe." *J Am Coll Nutr* 35 (2): 185–192. https://doi.org/10.1080/07315724.2015.1067872.
2. Hoffman MD, Bross TL, Hamilton RT. 2016. "Are we being drowned by overhydration advice on the Internet?" *Phys Sportsmed* 44 (4): 343–348. https://doi.org/10.1080/00913847.2016.1222853.
3. https://bostonmarathoncoach.wordpress.com/2011/04/28/remembering-dr-cynthia-lucero/.
4. Convertino VA, Armstrong LE, Coyle EF, et al. 1996. "American College of Sports Medicine position stand. Exercise and fluid replacement." *Med Sci Sports Exerc* 28 (1): i-vii. https://doi.org/10.1097/00005768-199610000-00045.
5. Almond CSD, Shin AY, Fortescue EB, et al. 2005. "Hyponatremia among Runners in the Boston Marathon." *N Engl J Med* 352 (15): 1550–1556. https://doi.org/10.1056/NEJMoa043901.
6. Church A, Lee F, Buono MJ. 2017. "Transition duration of ingested deuterium oxide to eccrine sweat during exercise in the heat." *J Therm Biol* 63: 88–91. https://doi.org/10.1016/j.jtherbio.2016.11.018.
7. Adolph EF. 1947. *Physiology of Man in the Desert*. 1st ed. New York: Interscience Publishing Co, 116.
8. Noakes T. 2012. *Waterlogged: The Serious Problem of Overhydration in Endurance Sports*. Champaign: Human Kinetics.
9. Montain SJ, Coyle EF. 1992. "Influence of graded dehydration on hyperthermia and cardiovascular drift during exercise." *J Appl Physiol* (1985) 73 (4): 1340–1350. https://doi.org/10.1152/jappl.1992.73.4.1340.
10. Adolph EF. 1947. *Physiology of Man in the Desert*. See Chapters 12, 13, and 14.
11. DeGroot D. Presentation by Col. David DeGroot, PhD. presented at: American College of Sports Medicine Annual Meeting; June 2023; Denver.
12. Speedy DB, Noakes T. 1999. "Hyponatremia in ultradistance triathletes." *Medicine & Science in Sports & Exercise* 31 (6): 809–815. https://journals.lww.com/acsm-msse/Fulltext/1999/06000/Hyponatremia_in_ultradistance_triathletes.8.aspx.
13. Wall BA, Watson G, Peiffer JJ, Abbiss CR, Siegel R, Laursen PB. 2015. "Current hydration guidelines are erroneous: dehydration does not impair exercise performance in the heat." *Br J Sports Med* 49 (16): 1077–1083. https://doi.org/10.1136/bjsports-2013-092417.
14. Barry, Samantha. "Brooke Shields Is in Her F-ck-It Era." *Glamour*, November 1, 2023.
15. Martin, Saleen. "Indiana mom dies at 35 from drinking too much water: What to know about water toxicity." *USA Today*, August 8, 2023 https://www.usatoday.com/story/life/health-wellness/2023/08/08/how-much-water-is-too-much-water-toxicity/70549738007/.

16. ABCNews, *Good Morning America*. "Jury Rules Against Radio Station After Water-Drinking Contest Kills Calif. Mom." November 1, 2009.
17. Schroeder, Roberta. "Crystal Water Bottles Are The Wellness Trend We Can Get Behind." *ELLE*, May 30, 2022. https://www.elle.com/uk/beauty/body-and-physical-health/a28242635/crystal-water-bottle-wellness-trend.
18. Watson P, Whale A, Mears SA, Reyner LA, Maughan RJ. 2015. "Mild hypohydration increases the frequency of driver errors during a prolonged, monotonous driving task." *Physiology & Behavior* 147: 313–318. https://doi.org/10.1016/j.physbeh.2015.04.028.
19. Speedy DB, Noakes TD, Schneider C. 2001. "Exercise-associated hyponatremia: A review." *Emergency Medicine* 13 (1): 17–27. https://doi.org/10.1046/j.1442-2026.2001.00173.x.
20. Valtin H. 2002. " 'Drink at least eight glasses of water a day.' Really? Is there scientific evidence for '8 x 8'?" *Am J Physiol Regul Integr Comp Physiol* 283 (5): R993–1004. https://doi.org/10.1152/ajpregu.00365.2002.
21. Yamada Y, Zhang X, Henderson MET, et al. 2022. "Variation in human water turnover associated with environmental and lifestyle factors." *Science* 378 (6622): 909–915. https://doi.org/10.1126/science.abm8668.
22. McDermott BP, Anderson SA, Armstrong LE, et al. 2017. "National Athletic Trainers' Association Position Statement: Fluid Replacement for the Physically Active." *Journal of Athletic Training* 52 (9): 877–895. https://doi.org/10.4085/1062-6050-52.9.02; American College of Sports M, Sawka MN, Burke LM, et al. 2007. "American College of Sports Medicine position stand. Exercise and fluid replacement." *Med Sci Sports Exerc* 39 (2): 377–90. https://doi.org/10.1249/mss.0b013e31802ca597; Goulet ED. 2012. "Dehydration and endurance performance in competitive athletes." *Nutr Rev* 70 Suppl 2: S132–6. https://doi.org/10.1111/j.1753-4887.2012.00530.x; Hoffman MD, Stellingwerff T, Costa RJS. 2019. "Considerations for ultra-endurance activities: part 2 - hydration." *Res Sports Med* 27 (2): 182–194. https://doi.org/10.1080/15438627.2018.1502189.
23. Cheuvront SN, Kenefick RW. 2022. "Personalized Hydration Requirements of Runners." *Int J Sport Nutr Exerc Metab* 32 (4): 233–237. https://doi.org/10.1123/ijsnem.2022-0001.
24. Cheuvront SN, Kenefick RW. 2021. "Personalized fluid and fuel intake for performance optimization in the heat." *J Sci Med Sport* 24 (8): 735–738. https://doi.org/10.1016/j.jsams.2021.01.004.
25. Maughan RJ, Watson P, Cordery PA, et al. 2016. "A randomized trial to assess the potential of different beverages to affect hydration status: development of a beverage hydration index." *American Journal of Clinical Nutrition* 103 (3): 717–723. https://doi.org/10.3945/ajcn.115.114769.
26. Grandjean AC, Reimers KJ, Bannick KE, Haven MC. 2000. "The effect of caffeinated, non-caffeinated, caloric and non-caloric beverages on hydration." *J Am Coll Nutr* 19 (5): 591–600. https://doi.org/10.1080/07315724.2000.10718956; Maughan RJ, Griffin J. 2003. "Caffeine ingestion and fluid balance: a review." *J Human Nutrition Diet* 16 (6): 411–420. https://doi.org/10.1046/j.1365-277X.2003.00477.x.

27. Naulleau C, Jeker D, Pancrate T, et al. 2022. "Effect of Pre-Exercise Caffeine Intake on Endurance Performance and Core Temperature Regulation During Exercise in the Heat: A Systematic Review with Meta-Analysis." *Sports Med* 52 (10): 2431–2445. https://doi.org/10.1007/s40279-022-01692-1.
28. Morris NB, Bain AR, Cramer MN, Jay O. 2014. "Evidence that transient changes in sudomotor output with cold and warm fluid ingestion are independently modulated by abdominal, but not oral thermoreceptors." *J Appl Physiol* (1985) 116 (8): 1088–95. https://doi.org/10.1152/japplphysiol.01059.2013.
29. Wijnen AHC, Steennis J, Catoire M, Wardenaar FC, Mensink M. 2016. "Post-Exercise Rehydration: Effect of Consumption of Beer with Varying Alcohol Content on Fluid Balance after Mild Dehydration." *Front Nutr* 3: 45. https://doi.org/10.3389/fnut.2016.00045.
30. Desbrow B, Murray D, Leveritt M. 2013. "Beer as a Sports Drink? Manipulating Beer's Ingredients to Replace Lost Fluid." *International Journal of Sport Nutrition and Exercise Metabolism* 23 (6): 593–600. https://doi.org/10.1123/ijsnem.23.6.593.

8. THE NICEST KID YOU'D EVER WANT TO MEET

1. WSB-TV 2 Atlanta. Student Athlete Dies of Heat Stroke. Atlanta: WSB-TV; 2011; "Football player from Georgia dies at camp." Staff Report. *Ocala Star Banner*. Ocala, FL. August 2, 2011.
2. NCAA Sport Science Institute. 2019. Preventing Catastrophic Injury and Death in Collegiate Athletes. *NCAA Sport Science Institute* (Indianapolis). https://ncaaorg.s3.amazonaws.com/ssi/injury_prev/SSI_PreventingCatastrophicInjury Booklet.pdf.
3. Grundstein AJ, Ramseyer C, Zhao F, et al. 2012. "A retrospective analysis of American football hyperthermia deaths in the United States." *Int J Biometeorol* 56 (1): 11–20. https://doi.org/10.1007/s00484-010-0391-4.
4. Francis K, Feinstein R, Brasher J. 1991. "Heat illness in football players in Alabama." *Ala Med* 60 (9): 10–14. http://www.ncbi.nlm.nih.gov/pubmed/2048546.
5. Kulka TJ, Kenney WL. 2002. "Heat balance limits in football uniforms how different uniform ensembles alter the equation." *Phys Sportsmed* 30 (7): 29–39. https://doi.org/10.3810/psm.2002.07.377.
6. Cooper ER, Grundstein AJ, Miles JD, et al. 2020. "Heat Policy Revision for Georgia High School Football Practices Based on Data-Driven Research." *Journal of Athletic Training* 55 (7): 673–681. https://doi.org/10.4085/1062-6050-542-18.
7. Cooper ER, Grundstein AJ, Miles JD, et al. 2020. "Heat Policy Revision for Georgia High School Football Practices Based on Data-Driven Research." *Journal of Athletic Training* 55 (7): 673–681. https://doi.org/10.4085/1062-6050-542-18.
8. Fieldstat, Elisha. Ring doorbell video shows UPS driver collapse in extreme Arizona heat. *NBCNews*. July 18, 2022. https://www.nbcnews.com/news/us-news/ring-video-shows-ups-driver-collapse-extreme-arizona-heat-rcna38663.

9. https://www.facebook.com/teamsters/posts/1132078408960959.
10. U.S. Bureau of Reclamation. The Story of Hoover Dam: Fatalities at Hoover Dam: 1931 and Earlier. https://www.usbr.gov/lc/hooverdam/history/essays/fat1931.html.
11. Chapman CL, Hess HW, Lucas RAI, et al. 2021. "Occupational heat exposure and the risk of chronic kidney disease of nontraditional origin in the United States." *Am J Physiol Regul Integr Comp Physiol* 321 (2): R141-R151. https://doi.org/10.1152/ajpregu.00103.2021.
12. Neely, Samantha, and Robiedo, Anthony. "As Texas swelters, local rules requiring water breaks for construction workers will soon be nullified." *USA Today*. April 16, 2024. https://www.usatoday.com/story/news/politics/2024/04/15/florida-removes-heat-protections-texas/73335597007/.
13. https://x.com/Burrows4TX/status/1623879154922885121.
14. Hansson E, Jakobsson K, Glaser J, et al. 2024. "Impact of heat and a rest-shade-hydration intervention program on productivity of piece-paid industrial agricultural workers at risk of chronic kidney disease of nontraditional origin." *Ann Work Expo Health* 68 (4): 366–375. https://doi.org/10.1093/annweh/wxae007.

9. TOO DARN HOT

1. Kulka TJ, Kenney WL. Heat balance limits in football uniforms how different uniform ensembles alter the equation. *Phys Sportsmed*. 2002;30(7):29–39. doi:10.3810/psm.2002.07.377.
2. Sherwood SC, Huber M. 2010. "An adaptability limit to climate change due to heat stress." *Proc Natl Acad Sci U S A* 107 (21): 9552–9555. https://doi.org/10.1073/pnas.0913352107.
3. Sherwood SC, Huber M. 2010. "An adaptability limit to climate change due to heat stress." *Proc Natl Acad Sci U S A* 107 (21): 9552–9555. https://doi.org/10.1073/pnas.0913352107.
4. Wallace-Wells D. "The Uninhabitable Earth." *New York*, July 9, 2017. https://nymag.com/intelligencer/2017/07/climate-change-earth-too-hot-for-humans.html.
5. Vecellio DJ, Wolf ST, Cottle RM, Kenney WL. "Evaluating the 35°C wet-bulb temperature adaptability threshold for young, healthy subjects (PSU HEAT Project)." *J Appl Physiol (1985)*. 2022 Feb 1;132(2):340-345. doi: 10.1152/japplphysiol.00738.2021.
6. Cottle RM, Fisher KG, Leach OK, Wolf ST, Kenney WL. 2024. "Critical environmental core temperature limits and heart rate thresholds across the adult age span (PSU HEAT Project)." *Journal of Applied Physiology* 137 (1): 145–153. https://doi.org/10.1152/japplphysiol.00117.2024.
7. https://www.nationalgeographic.com/history/article/stone-age-burial-ground-sahara-archaeology.
8. Wallace DC. 2005. "A mitochondrial paradigm of metabolic and degenerative diseases, aging, and cancer: a dawn for evolutionary medicine." *Annu Rev Genet* 39: 359–407. https://doi.org/10.1146/annurev.genet.39.110304.095751; Fumagalli M, Moltke I, Grarup N, et al. 2015. "Greenlandic Inuit show genetic signatures of diet and climate adaptation." *Science* 349 (6254): 1343–7. https://doi.org/10.1126/science.aab2319.

9. Alahmad B, Tobias A, Masselot P, Gasparrini A. 2025. "Are there more cold deaths than heat deaths?" *The Lancet Planetary Health* 9 (3): e170-e171. https://doi.org/10.1016/S2542-5196(25)00054-3; Gasparrini A, Guo Y, Hashizume M, et al. 2015. "Mortality risk attributable to high and low ambient temperature: a multicountry observational study." *Lancet* 386 (9991): 369–375. https://doi.org/10.1016/S0140-6736(14)62114-0; Zhao Q, Guo Y, Ye T, et al. 2021. "Global, regional, and national burden of mortality associated with non-optimal ambient temperatures from 2000 to 2019: a three-stage modelling study." *Lancet Planet Health* 5 (7): e415-e425. https://doi.org/10.1016/S2542-5196(21)00081-4.
10. Zhao Q, Guo Y, Ye T, et al. 2021. "Global, regional, and national burden of mortality associated with non-optimal ambient temperatures from 2000 to 2019: a three-stage modelling study." *Lancet Planet Health* 5 (7): e415-e425. https://doi.org/10.1016/S2542-5196(21)00081-4.
11. Wieners B. 2014. "No Way Is Matt Power Gone." *Bloomberg Businessweek*. https://www.bloomberg.com/news/articles/2014-04-11/no-way-is-matt-power-gone.
12. Gasparrini A, Guo Y, Hashizume M, et al. 2015. "Mortality risk attributable to high and low ambient temperature: a multicountry observational study." *Lancet* 386 (9991): 369–375. https://doi.org/10.1016/S0140-6736(14)62114-0.
13. Yu J, Ouyang Q, Zhu Y, Shen H, Cao G, Cui W. 2012. "A comparison of the thermal adaptability of people accustomed to air-conditioned environments and naturally ventilated environments." *Indoor Air* 22 (2): 110–118. https://doi.org/10.1111/j.1600-0668.2011.00746.x.
14. Nakamura K, Okada A, Watanabe H, et al. 2025. "In-hospital mortality of heat-related disease associated with wet bulb globe temperature: a Japanese nationwide inpatient data analysis." *Int J Biometeorol* 69 (4): 873–884. https://doi.org/10.1007/s00484-025-02867-x.
15. Kenney WL, Wolf ST, Dillon GA, Berry CW, Alexander LM. 2021. "Temperature regulation during exercise in the heat: Insights for the aging athlete." *Journal of Science and Medicine in Sport* 24 (8): 739–746. https://doi.org/10.1016/j.jsams.2020.12.007; Stapleton JM, Poirier MP, Flouris AD, et al. 2015. "Aging impairs heat loss, but when does it matter?" *J Appl Physiol* (1985) 118 (3): 299–309. https://doi.org/10.1152/japplphysiol.00722.2014.
16. Tankersley CG, Smolander J, Kenney WL, Fortney SM. 1991. "Sweating and skin blood flow during exercise: effects of age and maximal oxygen uptake." *J Appl Physiol* (1985) 71 (1): 236–242. https://doi.org/10.1152/jappl.1991.71.1.236.
17. Best, S., Thompson, M., Caillaud, C., Holvik, L., Fatseas, G., & Tammam, A. (2014). Exercise-heat acclimation in young and older trained cyclists. *Journal of Science and Medicine in Sport*, 17(6), 677–682. https://doi.org/10.1016/j.jsams.2013.10.243.
18. Tankersley CG, Smolander J, Kenney WL, Fortney SM. 1991. "Sweating and skin blood flow during exercise: effects of age and maximal oxygen uptake." *J Appl Physiol* (1985) 71 (1): 236–242. https://doi.org/10.1152/jappl.1991.71.1.236.
19. McIntyre RD, Zurawlew MJ, Oliver SJ, Cox AT, Mee JA, Walsh NP. 2021. "A comparison of heat acclimation by post-exercise hot water immersion and exercise

in the heat." *Journal of Science and Medicine in Sport* 24 (8): 729–734. https://doi.org/10.1016/j.jsams.2021.05.008.
20. Cole E., Donnan K. J., Simpson A. J., Garrett A. T. 2023. Short-term heat acclimation protocols for an aging population: Systematic review. *PloS one*, 18(3), e0282038. https://doi.org/10.1371/journal.pone.0282038.

10. STEAMED

1. Vahtla A. Finns' Friday night sauna in part behind day's peak power prices in Estonia. ERR News. 2024. https://news.err.ee/1609213489/finns-friday-night-sauna-in-part-behind-day-s-peak-power-prices-in-estonia.
2. Wellbeing Research Centre at the University of Oxford. 2025. "World Happiness Report: Data Sharing." Accessed July 16. https://www.worldhappiness.report/data-sharing/.
3. University of Eastern Finland. "Kuopio Ischaemic Heart Disease Risk Factor (KIHD) Study, Nutrition." University of Eastern Finland. Accessed July 10. https://uefconnect.uef.fi/en/kuopio-ischaemic-heart-disease-risk-factor-kihd-study-nutrition/#publications.
4. Quintela, Marco & Santos-Estévez, Manuel. (2015). Iron Age Saunas of Northern Portugal: State of the Art and Research Perspectives. *Oxford Journal of Archaeology*. 34. 10.1111/ojoa.12049.
5. Yegül FK. 2010. *Bathing in the Roman World*. 1. publ ed. Cambridge: Cambridge Univ. Press.
6. Garolla A, Torino M, Sartini B, et al. 2013. "Seminal and molecular evidence that sauna exposure affects human spermatogenesis." *Human Reproduction* 28 (4): 877–885. https://doi.org/10.1093/humrep/det020. https://academic.oup.com/humrep/article-lookup/doi/10.1093/humrep/det020.
7. Aaland M. 1978. *Sweat: the illustrated history and description of the Finnish sauna, Russian bania, Islamic hammam, Japanese mushi-buro, Mexican temescal, and American Indian & Eskimo sweat lodge*. Santa Barbara: Capra Press.
8. Viherjuuri H. 1972. *Sauna: The Finnish Bath*.

11. DETOXIFIED

1. Shakespeare J. 2006. "Bend it like Bikram." *The Guardian*. June 11, 2006.
2. Hunter SD, Laosiripisan J, Elmenshawy A, Tanaka H. 2018. "Effects of yoga interventions practised in heated and thermoneutral conditions on endothelium-dependent vasodilatation: The Bikram yoga heart study." *Experimental Physiology* 103 (3): 391–396. https://doi.org/10.1113/EP086725.
3. Lambert BS, Miller KE, Delgado DA, et al. 2020. "Acute Physiologic Effects of Performing Yoga in The Heat on Energy Expenditure, Range of Motion, and Inflammatory Biomarkers." *Int J Exerc Sci* 13 (3): 802–817. http://www.ncbi.nlm.nih.gov/pubmed/32509120.
4. Bourbeau KC, Moriarty TA, Bellovary BN, et al. 2021. "Cardiovascular, Cellular, and Neural Adaptations to Hot Yoga versus Normal-Temperature Yoga."

Int J Yoga 14 (2): 115–126. https://doi.org/10.4103/ijoy.IJOY_134_20; Hewett ZL, Cheema BS, Pumpa KL, Smith CA. 2015. "The Effects of Bikram Yoga on Health: Critical Review and Clinical Trial Recommendations." *Evid Based Complement Alternat Med* 2015: 428427. https://doi.org/10.1155/2015/428427; Hunter SD, Dhindsa M, Cunningham E, Tarumi T, Alkatan M, Tanaka H. 2013. "Improvements in glucose tolerance with Bikram Yoga in older obese adults: a pilot study." *J Bodyw Mov Ther* 17 (4): 404–407. https://doi.org/10.1016/j.jbmt.2013.01.002.

5. Perrotta AS, White MD, Koehle MS, Taunton JE, Warburton DER. 2018. "Efficacy of Hot Yoga as a Heat Stress Technique for Enhancing Plasma Volume and Cardiovascular Performance in Elite Female Field Hockey Players." *J Strength Cond Res* 32 (10): 2878–2887. https://doi.org/10.1519/JSC.0000000000002705.
6. Schnare DW, Denk G, Shields M, Brunton S. 1982. "Evaluation of a detoxification regimen for fat stored xenobiotics." *Medical Hypotheses* 9 (3): 265–282. https://doi.org/10.1016/0306-9877(82)90156-6.
7. Michelle O'Donnell. "Scientologist's Treatments Lure Firefighters." *The New York Times*. October 4, 2003.
8. Thomas Fuller. "Agent Orange Victims Get Scientology Treatment." *The New York Times*. September 5, 2012.
9. al-Zaki T, Jolly BT. 1997. "Severe hyponatremia after 'purification.'" *Ann Emerg Med* 29 (1): 194–195. https://doi.org/10.1016/s0196-0644(97)70335-4.
10. Baker LB. 2019. "Physiology of sweat gland function: The roles of sweating and sweat composition in human health." *Temperature* 6 (3): 211–259. https://doi.org/10.1080/23328940.2019.1632145; Baker LB, Wolfe AS. 2020. "Physiological mechanisms determining eccrine sweat composition." *Eur J Appl Physiol* 120 (4): 719–752. https://doi.org/10.1007/s00421-020-04323-7.
11. Baker and Wolfe. "Physiological mechanisms determining eccrine sweat composition."
12. Gambelunghe C, Rossi R, Aroni K, et al. 2013. "Sweat testing to monitor drug exposure." *Ann Clin Lab Sci* 43 (1): 22–30. http://www.ncbi.nlm.nih.gov/pubmed/23462602; Brasier N, Eckstein J. 2019. "Sweat as a Source of Next-Generation Digital Biomarkers." *Digit Biomark* 3 (3): 155–165. https://doi.org/10.1159/000504387.
13. Everts S. 2021. *The Joy of Sweat*. New York: W. W. Norton & Company.
14. Baker LB. 2019. "Physiology of sweat gland function."
15. Genuis SJ, Birkholz D, Rodushkin I, Beesoon S. 2011. "Blood, Urine, and Sweat (BUS) Study: Monitoring and Elimination of Bioaccumulated Toxic Elements." *Arch Environ Contam Toxicol* 61 (2): 344–357. https://doi.org/10.1007/s00244-010-9611-5.
16. Gordon CJ. 2003. Role of environmental stress in the physiological response to chemical toxicants. *Environmental Research*, 92(1), 1–7. https://doi.org/10.1016/s0013-9351(02)00008-7.
17. International Association of Firefighters. 2017. Sauna Use for Detoxification After Fire Suppression. https://share.ansi.org/PUBHDSSC/All%20HDSSC%20Public%20Documents/ANSI-HDSSC%20IAB%20Roundtable/2018%20Meeting/IAFF%20Sauna%20Information%202017%2008%2009.pdf.

12. THE CASE AGAINST COLD PLUNGING

1. Stocks JM, Taylor NA, Tipton MJ, Greenleaf JE. 2004. "Human physiological responses to cold exposure." *Aviat Space Environ* Med 75 (5): 444–457. https://www.ncbi.nlm.nih.gov/pubmed/15152898.
2. Weill Cornell Medical College. "The Rise and Decline of Psychiatric Hydrotherapy." http://psych-history.weill.cornell.edu/osk_die_lib/hydrotherapy/default.htm. Accessed June 4, 2024. Weill Cornell Medicine DeWitt Wallace Institute of Psychiatry: History, Policy, and the Arts.
3. Katie Tarrant and Keiran Southern. "Wim Hof started a cold water therapy trend. Our daughters died trying it." *The Sunday Times* (London). June 23, 2024. https://www.thetimes.com/uk/society/article/wim-hof-iceman-breathing-method-cold-water-therapy-killed-9tnvc5w8q.
4. Carney S, Hof W. 2017. *What Doesn't Kill Us: How Freezing Water, Extreme Altitude, and Environmental Conditioning Will Renew Our Lost Evolutionary Strength.* New York: Rodale Books.
5. Shattock MJ, Tipton MJ. 2012. "'Autonomic conflict': a different way to die during cold water immersion?" *The Journal of Physiology* 590 (14): 3219–3230. https://doi.org/10.1113/jphysiol.2012.229864.
6. Muzik O, Reilly KT, Diwadkar VA. 2018. "Brain over body"-A study on the willful regulation of autonomic function during cold exposure. *NeuroImage*, 172, 632–641. https://doi.org/10.1016/j.neuroimage.2018.01.06.7.
7. Betz MW, Fuchs CJ, Chedd F, et al. 2025. "Post-Exercise Cooling Lowers Skeletal Muscle Microvascular Perfusion and Blunts Amino Acid Incorporation into Muscle Tissue in Active Young Adults." *Medicine & Science in Sports & Exercise.* https://doi.org/10.1249/MSS.0000000000003723; Roberts LA, Raastad T, Markworth JF, et al. 2015. "Post-exercise cold water immersion attenuates acute anabolic signalling and long-term adaptations in muscle to strength training." *The Journal of Physiology* 593 (18): 4285–4301. https://doi.org/10.1113/JP270570.
8. Roberts LA, Raastad T, Markworth JF, et al. 2015. "Post-exercise cold water immersion attenuates acute anabolic signalling and long-term adaptations in muscle to strength training." *The Journal of Physiology* 593 (18): 4285–4301. https://doi.org/10.1113/JP270570.
9. Betz MW, Fuchs CJ, Chedd F, et al. 2025. "Post-Exercise Cooling Lowers Skeletal Muscle Microvascular Perfusion and Blunts Amino Acid Incorporation into Muscle Tissue in Active Young Adults." *Medicine & Science in Sports & Exercise.* https://doi.org/10.1249/MSS.0000000000003723.
10. Ahokas EK, Ihalainen JK, Hanstock HG, Savolainen E, Kyrolainen H. 2023. "A post-exercise infrared sauna session improves recovery of neuromuscular performance and muscle soreness after resistance exercise training." *Biol Sport* 40 (3): 681–689. https://doi.org/10.5114/biolsport.2023.119289.
11. Cheng AJ, Willis SJ, Zinner C, et al. 2017. "Post-exercise recovery of contractile function and endurance in humans and mice is accelerated by heating and slowed by cooling skeletal muscle." *J Physiol* 595 (24): 7413–7426. https://doi.org/10.1113/JP274870.
12. Dablainville V, Mornas A, Normand-Gravier T, et al. 2025. "Muscle regeneration

is improved by hot water immersion but unchanged by cold following a simulated musculoskeletal injury in humans." *J Physiol*. https://doi.org/10.1113/JP287777.
13. Hyldahl, Robert D et al. "Passive muscle heating attenuates the decline in vascular function caused by limb disuse." *The Journal of Physiology* vol. 599,20 (2021): 4581–4596. doi:10.1113/JP281900.
14. Mirkin G. Why Ice Delays Recovery. *DrMirkin* blog. 2021. https://drmirkin.com/fitness/why-ice-delays-recovery.html. Accessed August 20, 2025.
15. Malta ES, Dutra YM, Broatch JR, Bishop DJ, Zagatto AM. 2021. "The Effects of Regular Cold-Water Immersion Use on Training-Induced Changes in Strength and Endurance Performance: A Systematic Review with Meta-Analysis." *Sports Med* 51 (1): 161–174. https://doi.org/10.1007/s40279-020-01362-0.
16. Blades R, Mendes WB, Don BP, et al. 2024. "A randomized controlled clinical trial of a Wim Hof Method intervention in women with high depressive symptoms." *Compr Psychoneuroendocrinol* 20: 100272. https://doi.org/10.1016/j.cpnec.2024.100272.
17. Almahayni O, Hammond L. 2024. "Does the Wim Hof Method have a beneficial impact on physiological and psychological outcomes in healthy and non-healthy participants? A systematic review." *PLoS ONE* 19 (3): e0286933. https://doi.org/10.1371/journal.pone.0286933.
18. Briganti GL, Chesini G, Tarditi D, Serli D, Capodici A. 2023. "Effects of cold water exposure on stress, cardiovascular, and psychological variables." *Acta Physiologica* 239 (4): e14056. https://doi.org/10.1111/apha.14056.
19. Buijze GA, Sierevelt IN, van der Heijden BCJM, Dijkgraaf MG, Frings-Dresen MHW. 2016. "The Effect of Cold Showering on Health and Work: A Randomized Controlled Trial." *PLoS ONE* 11 (9): e0161749. https://doi.org/10.1371/journal.pone.0161749.
20. Oliver DM, McDougall CW, Robertson T, Grant B, Hanley N, Quilliam RS. 2023. "Self-reported benefits and risks of open water swimming to health, well-being and the environment: Cross-sectional evidence from a survey of Scottish swimmers." *PLoS ONE* 18 (8): e0290834. https://doi.org/10.1371/journal.pone.0290834.
21. Chang M, Ibaraki T, Naruse Y, Imamura Y. 2023. "A study on neural changes induced by sauna bathing: Neural basis of the "totonou" state." *PLoS ONE* 18 (11): e0294137. https://doi.org/10.1371/journal.pone.0294137.

13. THE HEAT CURE

1. Mason AE, Chowdhary A, Hartogensis W, et al. 2024. "Feasibility and acceptability of an integrated mind-body intervention for depression: whole-body hyperthermia (WBH) and cognitive behavioral therapy (CBT)." *International Journal of Hyperthermia* 41 (1): 2351459. https://doi.org/10.1080/02656736.2024.2351459.
2. Ward NG, Doerr HO, Storrie MC. 1983. "Skin conductance: a potentially sensitive test for depression." *Psychiatry Res* 10 (4): 295–302. https://doi.org/10.1016/0165-1781(83)90076-8.

3. Souetre E, Salvati E, Wehr TA, Sack DA, Krebs B, Darcourt G. 1988. "Twenty-four-hour profiles of body temperature and plasma TSH in bipolar patients during depression and during remission and in normal control subjects." *Am J Psychiatry* 145 (9): 1133–1137. https://doi.org/10.1176/ajp.145.9.1133.
4. Mason AE, Kasl P, Soltani S, et al. 2024. "Elevated body temperature is associated with depressive symptoms: results from the TemPredict Study." *Sci Rep* 14 (1): 1884. https://doi.org/10.1038/s41598-024-51567-w.
5. Laukkanen T, et al. 2018. "Sauna Bathing and Risk of Psychotic Disorders: A Prospective Cohort Study."
6. Srámek P, Simecková M, Janský L, Savlíková J, Vybíral S. 2000. "Human physiological responses to immersion into water of different temperatures." *Eur J Appl Physiol* 81 (5): 436–442. https://doi.org/10.1007/s004210050065. http://www.ncbi.nlm.nih.gov/pubmed/10751106.
7. Tipton MJ, Collier N, Massey H, Corbett J, Harper M. 2017. "Cold water immersion: kill or cure?" *Experimental Physiology* 102 (11): 1335–1355. https://doi.org/10.1113/EP086283.
8. CDC Newsroom. 2023. "Provisional Suicide Deaths in the United States, 2022." U.S. Centers for Disease Control and Prevention. Accessed July 17, 2024. https://www.cdc.gov/media/releases/2023/s0810-US-Suicide-Deaths-2022.html.
9. NHS Business Services Authority. 2024. "NHS Releases 2023/2024 mental health medicines statistics for England." Accessed 2025. https://media.nhsbsa.nhs.uk/press-releases/5171d616-95ea-4282-959b-15f8bfed6a0f/nhs-releases-2023-24-mental-health-medicines-statistics-for-england.
10. Organisation for Economic Co-operation and Development. 2025. "OECD Data Explorer: Pharmaceutical Consumption." Accessed July 16, 2025.
11. Kirsch I, Deacon BJ, Huedo-Medina TB, Scoboria A, Moore TJ, Johnson BT. 2008. "Initial severity and antidepressant benefits: a meta-analysis of data submitted to the Food and Drug Administration." *PLoS Med* 5 (2): e45. https://doi.org/10.1371/journal.pmed.0050045.
12. Nyer MB, Hopkins LB, Nagaswami M, et al. 2023. "A Randomized Controlled Trial of Community-Delivered Heated Hatha Yoga for Moderate-to-Severe Depression." *J Clin Psychiatry* 84 (6): 22m14621. https://doi.org/10.4088/JCP.22m14621.
13. Brinsley J, Schuch F, Lederman O, et al. 2021. "Effects of yoga on depressive symptoms in people with mental disorders: a systematic review and meta-analysis." *Br J Sports Med* 55 (17): 992–1000. https://doi.org/10.1136/bjsports-2019-101242; Cramer H, Anheyer D, Lauche R, Dobos G. 2017. "A systematic review of yoga for major depressive disorder." *Journal of Affective Disorders* 213: 70–77. https://doi.org/10.1016/j.jad.2017.02.006.
14. David-Néel A. 1932. *Magic and Mystery in Tibet*.
15. Benson H, Lehmann JW, Malhotra MS, Goldman RF, Hopkins J, Epstein MD. 1982. "Body temperature changes during the practice of g Tum-mo yoga." *Nature* 295 (5846): 234–236. https://doi.org/10.1038/295234a0.
16. Benson H, Klipper MZ. 2000. *The Relaxation Response*. New York: HarperTorch.
17. Benson H, Lehmann JW, Malhotra MS, Goldman RF, Hopkins J, Epstein MD. 1982. "Body temperature changes during the practice of g Tum-mo yoga." *Nature* 295 (5846): 234–236. https://doi.org/10.1038/295234a0.

18. Menard C, Hodes GE, Russo SJ. 2016. "Pathogenesis of depression: Insights from human and rodent studies." *Neuroscience* 321: 138–162. https://doi.org/10.1016/j.neuroscience.2015.05.053.
19. Nash D, Hughes MG, Butcher L, et al. 2023. "IL-6 signaling in acute exercise and chronic training: Potential consequences for health and athletic performance." *Scand J Med Sci Sports* 33 (1): 4–19. https://doi.org/10.1111/sms.14241.
20. Lowry CA, Hale MW, Evans AK, et al. 2008. "Serotonergic systems, anxiety, and affective disorder: focus on the dorsomedial part of the dorsal raphe nucleus." *Ann N Y Acad Sci* 1148: 86–94. https://doi.org/10.1196/annals.1410.004.
21. Janssen CW, Lowry CA, Mehl MR, et al. 2016. "Whole-Body Hyperthermia for the Treatment of Major Depressive Disorder: A Randomized Clinical Trial." *JAMA Psychiatry* 73 (8): 789. https://doi.org/10.1001/jamapsychiatry.2016.1031.
22. Heckel M. 1960. "Beliebig langdauernde und gezielt dosierbare Erhöhung der Körpertemperatur durch eine Infrarotbestrahlungsanordnung." Strahlentherapie 111 (1): 149–153. durch eine Infrarotbestrahlungsanordnung; Heckel-Reusser S. 2022. "Whole-Body Hyperthermia (WBH): Historical Aspects, Current Use, and Future Perspectives." In Water-filtered Infrared A (wIRA) Irradiation, edited by Vaupel P. Springer.
23. Engelking C. 2016. "Germ of an Idea: William Coley's Cancer-Killing Toxins." *Discover Magazine*. Accessed July 17. https://www.discovermagazine.com/health/germ-of-an-idea-william-coleys-cancer-killing-toxins.
24. Zschaeck S, Beck M. 2022. "Whole-Body Hyperthermia in Oncology: Renaissance in the Immunotherapy Era?" In Water-filtered Infrared A (wIRA) Irradiation, edited by Vaupel P, 107–115. Cham: Springer International Publishing.
25. Roberts JW, Powlovich L, Sheybani N, LeBlang S. 2022. "Focused ultrasound for the treatment of glioblastoma." J Neurooncol 157 (2): 237–247. https://doi.org/10.1007/s11060-022-03974-0.
26. Naumann J, Kruza I, Denkel L, Kienle G, Huber R. 2020. "Effects and feasibility of hyperthermic baths in comparison to exercise as add-on treatment to usual care in depression: a randomised, controlled pilot study." *BMC Psychiatry* 20 (1): 536. https://doi.org/10.1186/s12888-020-02941-1. https://bmcpsychiatry.biomedcentral.com/articles/10.1186/s12888-020-02941-1.
27. Williams LE, Bargh JA. 2008. "Experiencing Physical Warmth Promotes Interpersonal Warmth." *Science* 322 (5901): 606–607. https://doi.org/10.1126/science.1162548.
28. Inagaki TK, Eisenberger NI. 2013. "Shared neural mechanisms underlying social warmth and physical warmth." *Psychol Sci* 24 (11): 2272–2280. https://doi.org/10.1177/0956797613492773.
29. Williams and Bargh. 2008. "Experiencing Physical Warmth Promotes Interpersonal Warmth."
30. Quadt L, Critchley HD, Garfinkel SN. 2018. "The neurobiology of interoception in health and disease." *Ann N Y Acad Sci* 1428 (1): 112–128. https://doi.org/10.1111/nyas.13915. https://www.ncbi.nlm.nih.gov/pubmed/29974959.
31. Naumann J, Kruza I, Denkel L, Kienle G, Huber R. 2020. "Effects and feasibility of hyperthermic baths in comparison to exercise as add-on treatment to usual care in depression: a randomised, controlled pilot study." *BMC Psychiatry* 20

(1): 536. https://doi.org/10.1186/s12888-020-02941-1. https://bmcpsychiatry.biomedcentral.com/articles/10.1186/s12888-020-02941-1.

14. HOTWIRED

1. Essel E, Zavala EI, Schulz-Kornas E, et al. 2023. "Ancient human DNA recovered from a Palaeolithic pendant." *Nature* 618 (7964): 328–332. https://doi.org/10.1038/s41586-023-06035-2.
2. Lesté-Lasserre C. "DNA from 25,000-year-old tooth pendant reveals woman who wore it." *New Scientist*. May 3, 2023.
3. Morris J. 1998. *Manhattan '45*. Johns Hopkins Univ. Press.

BIBLIOGRAPHY

Aaland M. *Sweat: the illustrated history and description of the Finnish sauna, Russian bania, Islamic hammam, Japanese mushi-buro, Mexican temescal, and American Indian & Eskimo sweat lodge.* Capra Press; 1978:252.
Abdolhamidi S. *An ancient engineering feat that harnessed the wind.* BBC; 2018.
ACE Fitness. ACE-ProSource™: July 2013. ACE-sponsored Study: Hot Yoga—Go Ahead and Turn Up the Heat. https://www.acefitness.org/continuing-education/prosource/july-2013/3353/ace-sponsored-study-hot-yoga-go-ahead-and-turn-up-the-heat/.
ACE Fitness. ACE® Study Focuses on Safety of Bikram Yoga by Measuring Heart Rate and Core Temperatures During Class. https://www.acefitness.org/about-ace/press-room/press-releases/5388/ace-study-focuses-on-safety-of-bikram-yoga-by-measuring-heart-rate-and-core-temperatures-during-class/files/516/ace-study-focuses-on-safety-of-bikram-yoga-by-measuring-heart-rate-and-core-temperatures-during.html.
Adams EL, Casa DJ, Huggins RA, et al. Heat Exposure and Hypohydration Exacerbate Physiological Strain During Load Carrying. J Strength Cond Res. 2019;33(3):727–735. doi:10.1519/JSC.0000000000001831.
Adolph EF. *Physiology of Man in the Desert.* 1st ed. Interscience Publishing. 1947:371.
Ahokas EK, Ihalainen JK, Hanstock HG, Savolainen E, Kyrolainen H. A post-exercise infrared sauna session improves recovery of neuromuscular performance and muscle soreness after resistance exercise training. *Biol Sport.* 2023;40(3):681–689. doi:10.5114/biolsport.2023.119289.
al-Zaki T, Jolly BT. Severe hyponatremia after "purification." *Ann Emerg Med.* 1997;29(1):194–195. doi:10.1016/s0196-0644(97)70335-4.
Alahmad B, Tobias A, Masselot P, Gasparrini A. Are there more cold deaths than heat deaths? *The Lancet Planetary Health.* 2025;9(3):e170-e171. doi:10.1016/S2542-5196(25)00054-3.

Alcaldia de Medellin Secretaria de Medio Ambiente. Corredores y muros verdes. Accessed July 22, 2025. https://www.medellin.gov.co/es/secretaria-medio-ambiente/medellin-biodiversa/corredores-y-muros-verdes/.

Aldea D, Kamberov YG. En1 sweat we trust: How the evolution of an Engrailed 1 enhancer made humans the sweatiest ape. *Temperature (Austin)*. 2022;9(4):303–305. doi:10.1080/23328940.2021.2019548.

Allen WE, DeNardo LA, Chen MZ, et al. Thirst-associated preoptic neurons encode an aversive motivational drive. *Science*. 2017;357(6356):1149–1155. doi:10.1126/science.aan6747.

Almahayni O, Hammond L. Does the Wim Hof Method have a beneficial impact on physiological and psychological outcomes in healthy and non-healthy participants? A systematic review. PLoS ONE. 2024;19(3):e0286933. doi:10.1371/journal.pone.0286933.

Almond CSD, Shin AY, Fortescue EB, et al. Hyponatremia among Runners in the Boston Marathon. *N Engl J Med*. 2005;352(15):1550–1556. doi:10.1056/NEJMoa043901.

Alvarez L. Families of Athletes to Sue Over Heat-Related Deaths. *The New York Times*. 2012. https://www.nytimes.com/2012/08/01/us/families-to-sue-in-football-players-heat-related-deaths.html.

American College of Sports M, Sawka MN, Burke LM, et al. American College of Sports Medicine position stand. Exercise and fluid replacement. *Med Sci Sports Exerc*. 2007;39(2):377–90. doi:10.1249/mss.0b013e31802ca597.

Anderson DC. Opinion | by David C. Anderson. *The New York Times*. 1982/04/23/. Accessed June 27, 2025. https://www.nytimes.com/1982/04/23/opinion/by-david-c-anderson.html.

Anderson SA, Eichner ER, Bennett S, et al. Preventing Exertional Heat Stroke in Football: Time for a Paradigm Shift. *Sports Health*. 2024:19417381241260045. doi:10.1177/19417381241260045.

Andersson R. Promoter or enhancer, what's the difference? Deconstruction of established distinctions and presentation of a unifying model. *BioEssays*. 2015;37(3):314–323. doi:10.1002/bies.201400162.

Arai SR, Butzlaff A, Stotts NA, Puntillo KA. Quench the Thirst: Lessons From Clinical Thirst Trials. *Biological Research For Nursing*. 2014;16(4):456–466. doi:10.1177/1099800413505900.

Armstrong LE. Rehydration during Endurance Exercise: Challenges, Research, Options, Methods. *Nutrients*. 2021;13(3):887. doi:10.3390/nu13030887.

Aschwanden C. How Much Water Should I Drink? *The New York Times*. Accessed June 27, 2025. https://www.nytimes.com/2021/09/17/well/live/how-much-water-should-I-drink.html.

Baker LB. Physiology of sweat gland function: The roles of sweating and sweat composition in human health. *Temperature*. 2019;6(3):211–259. doi:10.1080/23328940.2019.1632145.

Baker LB, Wolfe AS. Physiological mechanisms determining eccrine sweat composition. *Eur J Appl Physiol*. 2020;120(4):719–752. doi:10.1007/s00421-020-04323-7.

Baranauskas MN, Constantini K, Paris HL, Wiggins CC, Schlader ZJ, Chapman RF. Heat Versus Altitude Training for Endurance Performance at Sea Level. *Exerc Sport Sci Rev*. 2021;49(1):50–58. doi:10.1249/JES.0000000000000238.

Barclay CJ, Curtin NA. The legacy of A. V. Hill's Nobel Prize winning work on muscle energetics. *The Journal of Physiology*. 2022;600(7):1555–1578. doi:10.1113/JP281556.

Barry H, Chaseling GK, Moreault S, et al. Improved neural control of body temperature following heat acclimation in humans. *The Journal of Physiology*. 2020;598(6):1223–1234. doi:10.1113/JP279266.

Barry S. Brooke Shields Is in Her F-ck-It Era. *Glamour*. 2023. https://www.glamour.com/story/brooke-shields-glamour-women-of-the-year-2023.

Behar R. Cover Story: The Thriving Cult of Greed and Power. *TIME* 2012.

Benson H, Klipper MZ. *The Relaxation Response*. HarperTorch; 2000:227.

Benson H, Lehmann JW, Malhotra MS, Goldman RF, Hopkins J, Epstein MD. Body temperature changes during the practice of g Tum-mo yoga. *Nature*. 1982; 295(5846):234–236. doi:10.1038/295234a0.

Benson L. About Utah: Running Man hits spotlight. *Deseret News*. 2009. https://www.deseret.com/2009/10/19/20347046/about-utah-running-man-hits-spotlight/

Best A, Lieberman DE, Kamilar JM. Diversity and evolution of human eccrine sweat gland density. *Journal of Thermal Biology*. 2019;84:331–338. doi:10.1016/j.jtherbio.2019.07.024.

Best, S., Thompson, M., Caillaud, C., Holvik, L., Fatseas, G., & Tammam, A. (2014). Exercise-heat acclimation in young and older trained cyclists. *Journal of science and medicine in sport*, 17(6), 677–682. https://doi.org/10.1016/j.jsams.2013.10.243.

Betz MW, Fuchs CJ, Chedd F, et al. Post-Exercise Cooling Lowers Skeletal Muscle Microvascular Perfusion and Blunts Amino Acid Incorporation into Muscle Tissue in Active Young Adults. *Medicine & Science in Sports & Exercise*. 2025. doi:10.1249/MSS.0000000000003723.

Blades R, Mendes WB, Don BP, et al. A randomized controlled clinical trial of a Wim Hof Method intervention in women with high depressive symptoms. *Compr Psychoneuroendocrinol*. 2024;20:100272. doi:10.1016/j.cpnec.2024.100272.

Blagden C. Experiments and Observations in an Heated Room. *Philosophical Transactions of the Royal Society*. 1775;1768–1775.

Bleakley C, McDonough S, Gardner E, Baxter GD, Hopkins JT, Davison GW. Cold-water immersion (cryotherapy) for preventing and treating muscle soreness after exercise. *Cochrane Database of Systematic Reviews*. 2012;2012(2). doi:10.1002/14651858.CD008262.pub2.

Bourbeau KC, Moriarty TA, Bellovary BN, et al. Cardiovascular, Cellular, and Neural Adaptations to Hot Yoga versus Normal-Temperature Yoga. *Int J Yoga*. 2021; 14(2):115–126. doi:10.4103/ijoy.IJOY_134_20.

Bramble DM, Lieberman DE. Endurance running and the evolution of Homo. *Nature*. 2004;432(7015):345–352. doi:10.1038/nature03052.

Brasier N, Eckstein J. Sweat as a Source of Next-Generation Digital Biomarkers. *Digit Biomark*. 2019;3(3):155–165. doi:10.1159/000504387.

Briganti GL, Chesini G, Tarditi D, Serli D, Capodici A. Effects of cold water exposure on stress, cardiovascular, and psychological variables. *Acta Physiol (Oxf)*. 2023;239(4):e14056. doi:10.1111/apha.14056.

Brinsley J, Schuch F, Lederman O, et al. Effects of yoga on depressive symptoms in people with mental disorders: a systematic review and meta-analysis. *Br J Sports Med*. 2021;55(17):992–1000. doi:10.1136/bjsports-2019-101242.

Brown HA, Topham TH, Clark B, et al. Quantifying Exercise Heat Acclimatisation in Athletes and Military Personnel: A Systematic Review and Meta-analysis. *Sports Med.* 2024;54(3):727–741. doi:10.1007/s40279-023-01972-4.

Brown P. Human Deaths from Hot and Cold Temperatures and Implications for Climate Change. *The Breakthrough Journal.* Dec. 1, 2022. https://thebreakthrough.org/issues/energy/human-deaths-from-hot-and-cold-temperatures-and-implications-for-climate-change.

Brunt VE, Howard MJ, Francisco MA, Ely BR, Minson CT. Passive heat therapy improves endothelial function, arterial stiffness and blood pressure in sedentary humans. *The Journal of Physiology.* 2016;594(18):5329–5342. doi:10.1113/JP272453.

Brunt VE, Minson CT. Heat therapy: mechanistic underpinnings and applications to cardiovascular health. *J Appl Physiol* (1985). 2021;130(6):1684–1704. doi:10.1152/japplphysiol.00141.2020.

Buijze GA, Sierevelt IN, van der Heijden BCJM, Dijkgraaf MG, Frings-Dresen MHW. The Effect of Cold Showering on Health and Work: A Randomized Controlled Trial. *PLoS ONE.* 2016;11(9):e0161749. doi:10.1371/journal.pone.0161749.

Buono, M. J., Kolding, M., Leslie, E., Moreno, D., Norwood, S., Ordille, A., & Weller, R. (2018). Heat acclimation causes a linear decrease in sweat sodium ion concentration. *Journal of thermal biology*, 71, 237–240. https://doi.org/10.1016/j.jtherbio.2017.12.001.

Cain T, Brinsley J, Bennett H, Nelson M, Maher C, Singh B. Effects of cold-water immersion on health and wellbeing: A systematic review and meta-analysis. *PLoS ONE.* 2025;20(1):e0317615. doi:10.1371/journal.pone.0317615.

Caldwell AR, Oki K, Ward SM, et al. Impact of successive exertional heat injuries on thermoregulatory and systemic inflammatory responses in mice. *J Appl Physiol* (1985). Nov 1 2021;131(5):1469–1485. doi:10.1152/japplphysiol.00160.2021.

calozada. It's Cold As Hell. *Witnessing Medieval Evil* blog. 2022. https://voices.uchicago.edu/witnessingmedievalevil/2022/03/13/its-cold-as-hell/.

Carey A. Clinic's results make 9/11 responders believe. Critics aside, they say Scientology's detox center cures ills. *The Philadelphia Inquirer.* October 7, 2007. Accessed Sept 1, 2024.

Carnegie Mellon University. Narconon Exposed: Does Narconon work? - Research Papers. https://www.cs.cmu.edu/~dst/Narconon/papers.htm.

Carney S, Hof W. *What Doesn't Kill Us: How Freezing Water, Extreme Altitude, and Environmental Conditioning Will Renew Our Lost Evolutionary Strength.* Rodale Books; 2017.

Carr AJ, Vallance BS, Rothwell J, Rea AE, Burke LM, Guy JH. Competing in Hot Conditions at the Tokyo Olympic Games: Preparation Strategies Used by Australian Race Walkers. *Front Physiol.* 2022;13:836858. doi:10.3389/fphys.2022.836858.

Carrier DR, Kapoor AK, Kimura T, et al. The Energetic Paradox of Human Running and Hominid Evolution [and Comments and Reply]. *Current Anthropology.* 1984 1984;25(4):483–495.

Carter R, 3rd, Cheuvront SN, Williams JO, et al. Epidemiology of hospitalizations and deaths from heat illness in soldiers. *Med Sci Sports Exerc.* 2005;37(8):1338–44. doi:10.1249/01.mss.0000174895.19639.ed.

Casa DJ, Stearns RL, Lopez RM, et al. Influence of hydration on physiological function and performance during trail running in the heat. *Journal of Athletic Training.* 2010;45(2):147–156. doi:10.4085/1062–6050–45.2.147.

CDC Newsroom. Provisional Suicide Deaths in the United States, 2022. U.S. Centers for Disease Control and Prevention. Accessed July 17, 2025. https://www.cdc.gov/media/releases/2023/s0810-US-Suicide-Deaths-2022.html.

Chang C-H, Bernard TE, Logan J. Effects of heat stress on risk perceptions and risk taking. Appl Ergon. 2017;62:150–157. doi:10.1016/j.apergo.2017.02.018.

Chapman CL, Hess HW, Lucas RAI, et al. Occupational heat exposure and the risk of chronic kidney disease of nontraditional origin in the United States. *Am J Physiol Regul Integr Comp Physiol.* 2021;321(2):R141-R151. doi:10.1152/ajpregu.00103.2021.

Cheng AJ, Willis SJ, Zinner C, et al. Post-exercise recovery of contractile function and endurance in humans and mice is accelerated by heating and slowed by cooling skeletal muscle. *J Physiol.* 2017;595(24):7413–7426. doi:10.1113/JP274870.

Cheuvront SN, Kenefick RW. Am I Drinking Enough? Yes, No, and Maybe. *J Am Coll Nutr.* 2016;35(2):185–192. doi:10.1080/07315724.2015.1067872.

Cheuvront SN, Kenefick RW. Personalized fluid and fuel intake for performance optimization in the heat. *J Sci Med Sport.* 24(8):735–738. doi:10.1016/j.jsams.2021.01.004.

Cheuvront SN, Kenefick RW. Personalized Hydration Requirements of Runners. *Int J Sport Nutr Exerc Metab.* 2022;32(4):233–237. doi:10.1123/ijsnem.2022-0001.

Convertino VA, Armstrong LE, Coyle EF, et al. American College of Sports Medicine position stand. Exercise and fluid replacement. *Med Sci Sports Exerc.* 1996;28(1):i-vii. doi:10.1097/00005768–199610000–00045.

Cooper ER, Grundstein AJ, Miles JD, et al. Heat Policy Revision for Georgia High School Football Practices Based on Data-Driven Research. *Journal of Athletic Training.* 2020;55(7):673–681. doi:10.4085/1062–6050–542–18.

Cotter JD, Thornton SN, Lee JK, Laursen PB. Are we being drowned in hydration advice? Thirsty for more? *Extrem Physiol Med.* 2014;3:18. doi:10.1186/2046–7648-3-18.

Cottle RM, Fisher KG, Leach OK, Wolf ST, Kenney WL. Critical environmental core temperature limits and heart rate thresholds across the adult age span (PSU HEAT Project). *Journal of Applied Physiology.* 2024;137(1):145–153. doi:10.1152/japplphysiol.00117.2024.

Cottle RM, Fisher KG, Wolf ST, Kenney WL. Onset of cardiovascular drift during progressive heat stress in young adults (PSU HEAT project). *Journal of Applied Physiology.* 2023;135(2):292–299. doi:10.1152/japplphysiol.00222.2023.

Cramer H, Anheyer D, Lauche R, Dobos G. A systematic review of yoga for major depressive disorder. *Journal of Affective Disorders.* 2017;213:70–77. doi:10.1016/j.jad.2017.02.006.

D'Angelo Friedman J. Study Finds Bikram Yoga Raises Body Temps to 103°+. *Yoga Journal.* 2021. https://www.yogajournal.com/lifestyle/bikram-yoga-causes-103-body-temps-study-finds-stay-safe/.

Dablainville V, Mornas A, Normand-Gravier T, et al. Muscle regeneration is improved by hot water immersion but unchanged by cold following a simulated musculoskeletal injury in humans. *J Physiol.* 2025. doi:10.1113/JP287777.

David-Néel A. *Magic and Mystery in Tibet.* 1932.

DeGroot D. Presentation by Col. David DeGroot, PhD. American College of Sports Medicine Annual Meeting; June 2023; Denver, CO.

Desbrow B, Murray D, Leveritt M. Beer as a Sports Drink? Manipulating Beer's Ingredients to Replace Lost Fluid. *International Journal of Sport Nutrition and Exercise Metabolism.* 2013;23(6):593–600. doi:10.1123/ijsnem.23.6.593.

Dixon PG, Mote TL. Patterns and Causes of Atlanta's Urban Heat Island-Initiated Precipitation. *Journal of Applied Meteorology and Climataolgy.* 2003;42(9):1273–1294. https://doi.org/10.1175/1520-0450(2003)042<1273:PACOAU>2.0.CO;2.

DJ Case and Associates. *The Nature of Americans.* 2025. https://natureofamericans.org

Donnan KJ, Williams EL, Bargh MJ. The effectiveness of heat preparation and alleviation strategies for cognitive performance: A systematic review. *Temperature (Austin).* 2023;10(4):404–433. doi:10.1080/23328940.2022.2157645.

Downs S. A Survey of Modern Life: Outdoor Time. *Medium* blog. 2022. https://medium.com/building-h/a-survey-of-modern-life-outdoor-time-3a99d9fa3acb.

Dreier F. David Roche Had Never Raced 100 Miles. He Still Smashed the Leadville 100 Ultramarathon. *Outside Online* blog. 2024. https://www.outsideonline.com/outdoor-adventure/hiking-and-backpacking/david-roche-leadville-100/.

Dunn EBF. Death For Millions in 1921's Record Heat Wave. *New York Herald.* 1921;61. Accessed Feb 26, 2025. https://www.newspapers.com/image/471536051/.

El Helou N, Tafflet M, Berthelot G, et al. Impact of Environmental Parameters on Marathon Running Performance. *PLoS ONE.* 2012;7(5):e37407. doi:10.1371/journal.pone.0037407.

Ely BR, Ely MR, Cheuvront SN, Kenefick RW, DeGroot DW, Montain SJ. Evidence against a 40°C core temperature threshold for fatigue in humans. *Journal of Applied Physiology.* 2009;107(5):1519–1525. doi:10.1152/japplphysiol.00577.2009.

Ely MR, Cheuvront SN, Roberts WO, Montain SJ. Impact of Weather on Marathon-Running Performance. *Medicine & Science in Sports & Exercise.* 2007;39(3):487–493. doi:10.1249/mss.0b013e31802d3aba.

Engelking C. Germ of an Idea: William Coley's Cancer-Killing Toxins. *Discover Magazine.* Accessed July 17, 2025. https://www.discovermagazine.com/health/germ-of-an-idea-william-coleys-cancer-killing-toxins.

Essel E, Zavala EI, Schulz-Kornas E, et al. Ancient human DNA recovered from a Palaeolithic pendant. *Nature.* 2023;618(7964):328–332. doi:10.1038/s41586-023-06035-2.

Everts S. *The Joy of Sweat.* W. W. Norton & Company; 2021:1.

Feder ME, Hofmann GE. Heat-shock proteins, molecular chaperones, and the stress response: evolutionary and ecological physiology. *Annu Rev Physiol.* 1999;61:243–82. doi:10.1146/annurev.physiol.61.1.243.

Ferriss T, Patrick R. Are Saunas the Next Big Performance-Enhancing "Drug"? *Tim Ferriss* blog. https://tim.blog/2014/04/10/saunas-hyperthermic-conditioning-2/.

Ferry D. There's a Serious Downside to All That Water You're Drinking. *Men's Health.* 2024.

Figaro MK, Mack GW. Regulation of fluid intake in dehydrated humans: role of oropharyngeal stimulation. *Am J Physiol.* 1997;272(6 Pt 2):R1740–1746. doi:10.1152/ajpregu.1997.272.6.R1740.

Fonseca F. Self-help guru convicted in sweat lodge deaths. Associated Press. 2011. https://www.nbcnews.com/news/amp/wbna43501833.

Foster H. Heart-attacks and the Sauna. *The Lancet.* 1976;308(7980):313. doi:10.1016/S0140-6736(76)90765-0.

Fotheringham W. *Put Me Back on My Bike: In Search of Tom Simpson.* Yellow Jersey Press; 2007:253.

Fox RH, Goldsmith R, Kidd DJ, Lewis HE. Acclimatization to heat in man by controlled elevation of body temperature. *J Physiol.* May 1963;166(3):530–47. doi:10.1113/jphysiol.1963.sp007121.

Francis K, Feinstein R, Brasher J. Heat illness in football players in Alabama. *Ala Med.* 1991;60(9):10–14.

Fuller T. Agent Orange Victims Get Scientology Treatment. *The New York Times.* 2012. Accessed Mar 27, 2025. https://www.nytimes.com/2012/09/06/world/asia/agent-orange-victims-in-vietnam-get-scientology-treatment.html.

Fumagalli M, Moltke I, Grarup N, et al. Greenlandic Inuit show genetic signatures of diet and climate adaptation. *Science.* 2015;349(6254):1343–7. doi:10.1126/science.aab2319.

Gambelunghe C, Rossi R, Aroni K, et al. Sweat testing to monitor drug exposure. *Ann Clin Lab Sci.* 2013;43(1):22–30.

Ganga, A., Jayaraman, M. V., E Santos Fontánez, S., Moldovan, K., Torabi, R., & Wolman, D. N. (2024). Population analysis of ischemic stroke burden and risk factors in the United States in the pre- and post-mechanical thrombectomy eras. *Journal of stroke and cerebrovascular diseases: the official journal of National Stroke Association, 33*(8), 107768. https://doi.org/10.1016/j.jstrokecerebrovasdis.2024.107768.

Garolla A, Torino M, Sartini B, et al. Seminal and molecular evidence that sauna exposure affects human spermatogenesis. *Human Reproduction.* 2013;28(4):877–885. doi:10.1093/humrep/det020.

Gasparrini A, Guo Y, Hashizume M, et al. Mortality risk attributable to high and low ambient temperature: a multicountry observational study. *Lancet.* 2015;386(9991):369–375. doi:10.1016/S0140-6736(14)62114-0.

Gay J. Chicken Coop Heat Lamps and 2:09 on a Treadmill: Meet the 'Mad Scientist' of Marathoning. *Wall Street Journal.* 2024.

Genuis SJ, Beesoon S, Birkholz D, Lobo RA. Human excretion of bisphenol A: blood, urine, and sweat (BUS) study. *J Environ Public Health.* 2012;2012:185731. doi:10.1155/2012/185731.

Genuis SJ, Beesoon S, Lobo RA, Birkholz D. Human elimination of phthalate compounds: blood, urine, and sweat (BUS) study. *ScientificWorldJournal.* 2012;2012:615068. doi:10.1100/2012/615068.

Genuis SJ, Birkholz D, Rodushkin I, Beesoon S. Blood, Urine, and Sweat (BUS) Study: Monitoring and Elimination of Bioaccumulated Toxic Elements. *Arch Environ Contam Toxicol.* 2011;61(2):344–357. doi:10.1007/s00244-010-9611-5.

Genuis SJ, Lane K, Birkholz D. Human Elimination of Organochlorine Pesticides: Blood, Urine, and Sweat Study. *Biomed Res Int.* 2016;2016:1624643. doi:10.1155/2016/1624643.

Genuis SK, Birkholz D, Genuis SJ. Human Excretion of Polybrominated Diphenyl Ether Flame Retardants: Blood, Urine, and Sweat Study. *Biomed Res Int.* 2017;2017:3676089. doi:10.1155/2017/3676089.

Gibson OR, Tuttle JA, Watt PW, Maxwell NS, Taylor L. Hsp72 and Hsp90alpha mRNA transcription is characterised by large, sustained changes in core temperature during heat acclimation. *Cell Stress and Chaperones.* 2016;21(6):1021–1035. doi:10.1007/s12192-016-0726-0.

Gifford B. How To Make Sweat Your Superpower. *Men's Health.* 2023.

Giuliani C, Peri A. Effects of Hyponatremia on the Brain. *J Clin Med.* 2014;3(4):1163–1177. doi:10.3390/jcm3041163.

Godek SF, Bartolozzi AR, Burkholder R, Sugarman E, Dorshimer G. Core temperature and percentage of dehydration in professional football linemen and backs during preseason practices. *Journal of Athletic Training.* 2006;41(1):8–14; discussion 14–17.

Gomez Isaza DF, Rodgers EM. Exercise training does not affect heat tolerance in Chinook salmon (Oncorhynchus tshawytscha). *Comp Biochem Physiol A Mol Integr Physiol.* 2022;270:111229. doi:10.1016/j.cbpa.2022.111229.

Goodell J. *The Heat Will Kill You First: Life and Death on a Scorched Planet.* Back Bay Books; 2024:416.

Goulet ED. Dehydration and endurance performance in competitive athletes. *Nutr Rev.* Nov 2012;70 Suppl 2:S132–6. doi:10.1111/j.1753–4887.2012.00530.x.

Grandjean AC, Reimers KJ, Bannick KE, Haven MC. The effect of caffeinated, non-caffeinated, caloric and non-caloric beverages on hydration. *J Am Coll Nutr.* 2000;19(5):591–600. doi:10.1080/07315724.2000.10718956.

Gruenling J. Coroners report for Monticello woman believed to have died from water toxicity released. WRTV News. 2023/08/02/. https://www.wrtv.com/news/public-safety/monticello-woman-dies-from-water-toxicity-her-family-is-raising-awareness.

Grundstein AJ, Ramseyer C, Zhao F, et al. A retrospective analysis of American football hyperthermia deaths in the United States. *Int J Biometeorol.* 2012;56(1):11–20. doi:10.1007/s00484-010-0391-4.

Hansson E, Jakobsson K, Glaser J, et al. Impact of heat and a rest-shade-hydration intervention program on productivity of piece-paid industrial agricultural workers at risk of chronic kidney disease of nontraditional origin. *Ann Work Expo Health.* 2024;68(4):366–375. doi:10.1093/annweh/wxae007.

Hanusch K-U, Janssen CH, Billheimer D, et al. Whole-Body Hyperthermia for the Treatment of Major Depression: Associations With Thermoregulatory Cooling. *AJP.* 2013;170(7):802–804. doi:10.1176/appi.ajp.2013.12111395.

Hari J. *Lost Connections: Uncovering the Real Causes of Depression—and the Unexpected Solutions.* Bloomsbury; 2018:321.

Heckel-Reusser S. Whole-Body Hyperthermia (WBH): Historical Aspects, Current Use, and Future Perspectives. In: Vaupel P, ed. *Water-filtered Infrared A (wIRA) Irradiation.* Springer; 2022.

Heming T. The Very Human Way Coach Olav Bu Builds Norwegian Triathlon Machines. *Triathlete.*

Hersh AM, Bhimreddy M, Weber-Levine C, et al. Applications of Focused Ultrasound for the Treatment of Glioblastoma: A New Frontier. *Cancers (Basel).* 2022;14(19). doi:10.3390/cancers14194920.

Hewett ZL, Cheema BS, Pumpa KL, Smith CA. The Effects of Bikram Yoga on Health: Critical Review and Clinical Trial Recommendations. *Evid Based Complement Alternat Med.* 2015;2015:428427. doi:10.1155/2015/428427.

Hill AB. The Environment and Disease: Association or Causation? *Proceedings of the Royal Society of Medicine.* 1965;58(5):295–300.doi:10.1177/003591576505800503.

Hoffman MD, Bross TL, Hamilton RT. Are we being drowned by overhydration advice on the Internet? *Phys Sportsmed.* 2016;44(4):343–348. doi:10.1080/00913847.2016.1222853.

Hoffman MD, Stellingwerff T, Costa RJS. Considerations for ultra-endurance activities: part 2 - hydration. *Res Sports Med.* 2019;27(2):182–194. doi:10.1080/15438627.2018.1502189.

Hosier G. How Much Does Heat Slow Down Your Race Pace? *Outside Online* blog. 2019. https://www.outsideonline.com/health/running/racing/race-strategy/how-much-does-heat-slow-your-race-pace/.

Hsieh M. Recommendations for treatment of hyponatraemia at endurance events. *Sports Med.* 2004;34(4):231–238. doi:10.2165/00007256-200434040-00003.

Hunter SD, Dhindsa M, Cunningham E, Tarumi T, Alkatan M, Tanaka H. Improvements in glucose tolerance with Bikram Yoga in older obese adults: a pilot study. *J Bodyw Mov Ther.* 2013;17(4):404–407. doi:10.1016/j.jbmt.2013.01.002.

Hunter SD, Dhindsa MS, Cunningham E, et al. Impact of Hot Yoga on Arterial Stiffness and Quality of Life in Normal and Overweight/Obese Adults. *J Phys Act Health.* 2016;13(12):1360–1363. doi:10.1123/jpah.2016-0170.

Hunter SD, Laosiripisan J, Elmenshawy A, Tanaka H. Effects of yoga interventions practised in heated and thermoneutral conditions on endothelium-dependent vasodilatation: The Bikram yoga heart study. *Experimental Physiology.* 2018;103(3):391–396. doi:10.1113/EP086725.

Hutchinson A. *Endure: mind, body, and the curiously elastic limits of human performance.* Revised and updated ed. Custom House; 2021:312.

Inagaki TK, Eisenberger NI. Shared neural mechanisms underlying social warmth and physical warmth. *Psychol Sci.* 2013;24(11):2272–80. doi:10.1177/0956797613492773.

International Association of Firefighters. Sauna Use for Detoxification After Fire Suppression. 2017.

International Bottled Water Association. Bottled Water Outsells Carbonated Soft Drinks for the Eighth Year in a Row. 2024. https://bottledwater.org/nr/bottled-water-outsells-carbonated-soft-drinks-for-the-eighth-year-in-a-row.

Ioannou LG, Tsoutsoubi L, Mantzios K, et al. Indicators to assess physiological heat strain–Part 3: Multi-country field evaluation and consensus recommendations. *Temperature.* 2022;9(3):274–291. doi:10.1080/23328940.2022.2044739.

IPCC. IPCC, 2023: Climate Change 2023: Synthesis Report. Contribution of Working Groups I, II and III to the Sixth Assessment Report of the Intergovernmental Panel on Climate Change [Core Writing Team, H. Lee and J. Romero (eds.)]. 2023:35–115. IPCC, Geneva, Switzerland.

Janssen CW, Lowry CA, Mehl MR, et al. Whole-Body Hyperthermia for the Treatment of Major Depressive Disorder: A Randomized Clinical Trial. *JAMA Psychiatry.* 2016;73(8):789. doi:10.1001/jamapsychiatry.2016.1031.

Jezova D, Vigas M, Tatar P, Jurcovicova J, Palat M. Rise in plasma beta-endorphin and ACTH in response to hyperthermia in sauna. *Horm Metab Res.* 1985;17(12):693–4. doi:10.1055/s-2007–1013648.

Kakamu T, Wada K, Smith DR, Endo S, Fukushima T. Preventing heat illness in the anticipated hot climate of the Tokyo 2020 Summer Olympic Games. *Environ Health Prev Med.* 2017;22(1):68. doi:10.1186/s12199-017-0675-y.

Kamberov YG, Guhan SM, DeMarchis A, et al. Comparative evidence for the independent evolution of hair and sweat gland traits in primates. *Journal of Human Evolution.* 2018;125:99–105. doi:10.1016/j.jhevol.2018.10.008.

Kamberov YG, Karlsson EK, Kamberova GL, et al. A genetic basis of variation in eccrine sweat gland and hair follicle density. *Proc Natl Acad Sci USA.* 2015;112(32):9932–9937. doi:10.1073/pnas.1511680112.

Keefe MS, Luk H-Y, Rolloque J-JS, Jiwan NC, McCollum TB, Sekiguchi Y. The weight, urine colour and thirst Venn diagram is an accurate tool compared with urinary and blood markers for hydration assessment at morning and afternoon timepoints in euhydrated and free-living individuals. *Br J Nutr.* 2024;131(7):1181–1188. doi:10.1017/S000711452300274X.

Kenney WL, Wolf ST, Dillon GA, Berry CW, Alexander LM. Temperature regulation during exercise in the heat: Insights for the aging athlete. *Journal of Science and Medicine in Sport.* 2021;24(8):739–746. doi:10.1016/j.jsams.2020.12.007

Kerr ZY, Register-Mihalik JK, Pryor RR, et al. The Association between Mandated Preseason Heat Acclimatization Guidelines and Exertional Heat Illness during Preseason High School American Football Practices. *Environ Health Perspect.* 2019;127(4):047003. doi:10.1289/EHP4163.

Kerr ZY, Scarneo-Miller SE, Yeargin SW, et al. Exertional Heat-Stroke Preparedness in High School Football by Region and State Mandate Presence. *Journal of Athletic Training.* 2019;54(9):921–928. doi:10.4085/1062–6050–581–18.

Kipps C, Sharma S, Tunstall Pedoe D. The incidence of exercise-associated hyponatraemia in the London marathon. *Br J Sports Med.* 2011;45(1):14–19. doi:10.1136/bjsm.2009.059535.

Kirsch I, Deacon BJ, Huedo-Medina TB, Scoboria A, Moore TJ, Johnson BT. Initial severity and antidepressant benefits: a meta-analysis of data submitted to the Food and Drug Administration. *PLoS Med.* 2008;5(2):e45. doi:10.1371/journal.pmed.0050045.

Kivimäki M, Virtanen M, Ferrie JE. The Link Between Sauna Bathing and Mortality May Be Noncausal. *JAMA Intern Med.* 2015;175(10):1718. doi:10.1001/jamainternmed.2015.3426.

Klingert M, Nikolaidis PT, Weiss K, Thuany M, Chlíbková D, Knechtle B. Exercise-Associated Hyponatremia in Marathon Runners. *J Clin Med.* 2022;11(22):6775. doi:10.3390/jcm11226775.

Knekt P, Järvinen R, Rissanen H, Heliövaara M, Aromaa A. Does sauna bathing protect against dementia? *Prev Med Rep.* 2020;20:101221. doi:10.1016/j.pmedr.2020.101221.

Kricho, Kade. China's Stone Age Skiers and History's Harsh Lessons. *The New York Times*, April 19, 2017. https://www.nytimes.com/2017/04/19/sports/skiing/skiing-china-cave-paintings.html.

Kulka TJ, Kenney WL. Heat balance limits in football uniforms how different uniform ensembles alter the equation. *Phys Sportsmed.* 2002;30(7):29–39. doi:10.3810/psm.2002.07.377.

Kunutsor SK, Khan H, Laukkanen T, Laukkanen JA. Joint associations of sauna bathing and cardiorespiratory fitness on cardiovascular and all-cause mortality risk: a long-term prospective cohort study. *Annals of Medicine.* 2018;50(2):139–146. doi:10.1080/07853890.2017.1387927.

Kunutsor, S. K., Khan, H., Zaccardi, F., Laukkanen, T., Willeit, P., & Laukkanen, J. A. (2018). Sauna bathing reduces the risk of stroke in Finnish men and women: A prospective cohort study. *Neurology, 90*(22), e1937–e1944. https://doi.org/10.1212/WNL.0000000000005606.

Kupperman KO. Fear of Hot Climates in the Anglo-American Colonial Experience. *The William and Mary Quarterly.* 1984;41(2):213. doi:10.2307/1919050.

Lambert BS, Miller KE, Delgado DA, et al. Acute Physiologic Effects of Performing Yoga in The Heat on Energy Expenditure, Range of Motion, and Inflammatory Biomarkers. *Int J Exerc Sci.* 2020;13(3):802–817.

Lasisi T, Smallcombe JW, Kenney WL, et al. Human scalp hair as a thermoregulatory adaptation. *Proc Natl Acad Sci USA.* 2023;120(24):e2301760120. doi:10.1073/pnas.2301760120.

Laskas JM. The Enlightened Man. *Esquire.* 2001.

Laukkanen JA, Laukkanen T, Kunutsor SK. Cardiovascular and Other Health Benefits of Sauna Bathing: A Review of the Evidence. *Mayo Clinic Proceedings.* 2018;93(8):1111–1121. doi:10.1016/j.mayocp.2018.04.008.

Laukkanen JA, Mäkikallio TH, Khan H, Laukkanen T, Kauhanen J, Kunutsor SK. Finnish sauna bathing does not increase or decrease the risk of cancer in men: A prospective cohort study. *Eur J Cancer.* 2019;121:184–191. doi:10.1016/j.ejca.2019.08.031.

Laukkanen T, Khan H, Zaccardi F, Laukkanen JA. Association Between Sauna Bathing and Fatal Cardiovascular and All-Cause Mortality Events. *JAMA Intern Med.* 2015;175(4):542. doi:10.1001/jamainternmed.2014.8187.

Laukkanen T, Kunutsor S, Kauhanen J, Laukkanen JA. Sauna bathing is inversely associated with dementia and Alzheimer's disease in middle-aged Finnish men. *Age and Ageing.* 2017;46(2):245–249. doi:10.1093/ageing/afw212.

Laukkanen T, Kunutsor SK, Khan H, Willeit P, Zaccardi F, Laukkanen JA. Sauna bathing is associated with reduced cardiovascular mortality and improves risk prediction in men and women: a prospective cohort study. *BMC Med.* 2018;16(1):219. doi:10.1186/s12916-018-1198-0.

Laukkanen T, Laukkanen JA, Kunutsor SK. Sauna Bathing and Risk of Psychotic Disorders: A Prospective Cohort Study. *Med Princ Pract.* 2018;27(6):562–569. doi:10.1159/000493392.

Laursen PB, Suriano R, Quod MJ, et al. Core temperature and hydration status during an Ironman triathlon. *Br J Sports Med.* 2006;40(4):320–325; discussion 325. doi:10.1136/bjsm.2005.022426.

Lee E, Kolunsarka I, Kostensalo J, et al. Effects of regular sauna bathing in conjunction with exercise on cardiovascular function: a multi-arm, randomized controlled trial. *Am J Physiol Regul Integr Comp Physiol.* 2022;323(3):R289-R299. doi:10.1152/ajpregu.00076.2022.

Leppäluoto J. Human thermoregulation in sauna. *Ann Clin Res*. 1988;20(4):240–243.

Leppäluoto J, Huttunen P, Hirvonen J, Väänänen A, Tuominen M, Vuori J. Endocrine effects of repeated sauna bathing. *Acta Physiologica Scandinavica*. 1986;128(3):467–470. doi:10.1111/j.1748-1716.1986.tb08000.x.

Lesté-Lasserre C. DNA from 25,000-year-old tooth pendant reveals woman who wore it. *New Scientist*. May 3, 2023.

Levick JJ. Remarks on sunstroke. *American Journal of Medicine*. 1859;37(43).

Lieberman DE. *The Story of the Human Body: Evolution, Health, and Disease*. Pantheon Books, 2013.

Lieberman DE. Human Locomotion and Heat Loss: An Evolutionary Perspective. In: Terjung R, ed. *Comprehensive Physiology*. 1 ed. Wiley; 2014:99–117.

Lieberman DE. *Exercised: Why Something We Never Evolved to Do Is Healthy and Rewarding*. Pantheon Books, 2020.

Liu S, Wen D, Feng C, et al. Alteration of gut microbiota after heat acclimation may reduce organ damage by regulating immune factors during heat stress. Front Microbiol. 2023;14:1114233. doi:10.3389/fmicb.2023.1114233.

Lopatin IA. Origin of the Native American Steam Bath. *American Anthropologist*. 1960;62(6):977–993.

Lorenzo S, Halliwill JR, Sawka MN, Minson CT. Heat acclimation improves exercise performance. *J Appl Physiol* (1985). 2010;109(4):1140–1147. doi:10.1152/japplphysiol.00495.2010.

Lowry CA, Hale MW, Evans AK, et al. Serotonergic systems, anxiety, and affective disorder: focus on the dorsomedial part of the dorsal raphe nucleus. *Ann N Y Acad Sci*. 2008;1148:86–94. doi:10.1196/annals.1410.004.

Lu C, Fuchs E. Sweat gland progenitors in development, homeostasis, and wound repair. *Cold Spring Harb Perspect Med*. 2014;4(2):a015222. doi:10.1101/cshperspect.a015222.

Lu Y-C, Romps DM. Extending the Heat Index. *Journal of Applied Meteorology and Climatology*. 2022;61(10):1367–1383. doi:10.1175/JAMC-D-22-0021.1.

Lundby C, Hamarsland H, Hansen J, et al. Hematological, skeletal muscle fiber, and exercise performance adaptations to heat training in elite female and male cyclists. *J Appl Physiol* (1985). 2023;135(1):217–226. doi:10.1152/japplphysiol.00115.2023.

Mahtheus Gouvea de Andrade. How Medellin is beating the heat with green corridors. BBC. Accessed July 17, 2025. https://www.bbc.com/future/article/20230922-how-medellin-is-beating-the-heat-with-green-corridors.

Malta ES, Dutra YM, Broatch JR, Bishop DJ, Zagatto AM. The Effects of Regular Cold-Water Immersion Use on Training-Induced Changes in Strength and Endurance Performance: A Systematic Review with Meta-Analysis. *Sports Med*. 2021;51(1):161–174. doi:10.1007/s40279-020-01362-0.

Mann R. Temperatures reached a deadly 48.8°C during the Hoover Dam construction. The Weather Network. Updated July 7, 2023. Accessed July 14, 2025. https://www.theweathernetwork.com/en/news/weather/severe/this-day-in-weather-history-july-7-1930-building-of-hoover-dam-construction.

Mantzios K, Ioannou LG, Panagiotaki Z, et al. Effects of Weather Parameters on Endurance Running Performance: Discipline-specific Analysis of 1258 Races. *Med Sci Sports Exerc*. 2022;54(1):153–161. doi:10.1249/MSS.0000000000002769.

Maron MB. The University of California Institute of Environmental Stress marathon field studies. *Advances in Physiology Education*. 2014;38(1):3–11. doi:10.1152/advan.00118.2013.

Maron MB, Wagner JA, Horvath SM. Thermoregulatory responses during competitive marathon running. *J Appl Physiol Respir Environ Exerc Physiol*. 1977;42(6):909–14. doi:10.1152/jappl.1977.42.6.909.

Martin S. How much water is too much? Indiana mom dies at 35 of water toxicity. *USA Today*. https://www.usatoday.com/story/life/health-wellness/2023/08/08/how-much-water-is-too-much-water-toxicity/70549738007/.

Mason AE, Chowdhary A, Hartogensis W, et al. Feasibility and acceptability of an integrated mind-body intervention for depression: whole-body hyperthermia (WBH) and cognitive behavioral therapy (CBT). *International Journal of Hyperthermia*. 2024;41(1):2351459. doi:10.1080/02656736.2024.2351459.

Mason AE, Kasl P, Soltani S, et al. Elevated body temperature is associated with depressive symptoms: results from the TemPredict Study. *Sci Rep*. 2024;14(1):1884. doi:10.1038/s41598-024-51567-w.

Maughan RJ, Griffin J. Caffeine ingestion and fluid balance: a review. *J Human Nutrition Diet*. 2003;16(6):411–420. doi:10.1046/j.1365-277X.2003.00477.x.

Maughan RJ, Watson P, Cordery PA, et al. A randomized trial to assess the potential of different beverages to affect hydration status: development of a beverage hydration index. *The American Journal of Clinical Nutrition*. 2016;103(3):717–723. doi:10.3945/ajcn.115.114769.

McCafferty C. Brooke Shields had a grand mal seizure–here's what you need to know about the condition. *The Conversation* blog. 2023. https://theconversation.com/brooke-shields-had-a-grand-mal-seizure-heres-what-you-need-to-know-about-the-condition-216962.

McDermott BP, Anderson SA, Armstrong LE, et al. National Athletic Trainers' Association Position Statement: Fluid Replacement for the Physically Active. *Journal of Athletic Training*. 2017;52(9):877–895. doi:10.4085/1062-6050-52.9.02.

McDougall C. *Born to Run: A Hidden Tribe, Superathletes, and the Greatest Race the World Has Never Seen*. Vintage; 2011:304.

McIntyre RD, Zurawlew MJ, Oliver SJ, Cox AT, Mee JA, Walsh NP. A comparison of heat acclimation by post-exercise hot water immersion and exercise in the heat. *Journal of Science and Medicine in Sport*. 2021;24(8):729–734. doi:10.1016/j.jsams.2021.05.008.

Menard C, Hodes GE, Russo SJ. Pathogenesis of depression: Insights from human and rodent studies. *Neuroscience*. 2016;321:138–162. doi:10.1016/j.neuroscience.2015.05.053.

Miller A. Before Air-Conditioning. *The New Yorker*. 1998.

Minsberg T. The Real Hurdle at Tokyo's Olympic Test Events: The Heat. *The New York Times*. Accessed June 27, 2025. https://www.nytimes.com/2019/08/15/sports/olympics/the-real-hurdle-at-tokyos-olympic-test-events-the-heat.html

Minson CT, Cotter JD. CrossTalk proposal: Heat acclimatization does improve performance in a cool condition. *Journal of Physiology*. 2016;594(2):241–243. doi:10.1113/JP270879.

Mirkin G. Why Ice Delays Recovery. *DrMirkin* blog. 2021. https://drmirkin.com/fitness/why-ice-delays-recovery.html.

Montain SJ, Coyle EF. Influence of graded dehydration on hyperthermia and cardiovascular drift during exercise. *J Appl Physiol* (1985). 1992;73(4):1340–1350. doi:10.1152/jappl.1992.73.4.1340.

Mooney C. Scientists challenge magazine story about 'uninhabitable Earth'. *Washington Post*. 2017. https://www.washingtonpost.com/news/energy-environment/wp/2017/07/12/scientists-challenge-magazine-story-about-uninhabitable-earth/

Moran DS, Erlich T, Epstein Y. The Heat Tolerance Test: An Efficient Screening Tool for Evaluating Susceptibility to Heat. 2007. doi:10.1123/jsr.16.3.215.

Morin E, Winterhalder B. Ethnography and ethnohistory support the efficiency of hunting through endurance running in humans. *Nat Hum Behav*. 2024;8(6):1065–1075. doi:10.1038/s41562-024-01876-x.

Morris NB, Bain AR, Cramer MN, Jay O. Evidence that transient changes in sudomotor output with cold and warm fluid ingestion are independently modulated by abdominal, but not oral thermoreceptors. *J Appl Physiol* (1985). 2014;116(8):1088–95. doi:10.1152/japplphysiol.01059.2013.

Morse MS. *Chilly Reception*. Smithsonian. 2002.

Moussaieff A, Rimmerman N, Bregman T, et al. Incensole acetate, an incense component, elicits psychoactivity by activating TRPV3 channels in the brain. *FASEB J*. 2008;22(8):3024–34. doi:10.1096/fj.07–101865.

Munsters C, Siegers E, Sloet van Oldruitenborgh-Oosterbaan M. Effect of a 14-Day Period of Heat Acclimation on Horses Using Heated Indoor Arenas in Preparation for Tokyo Olympic Games. *Animals (Basel)*. 2024;14(4):546. doi:10.3390/ani14040546.

Murray R. Dehydration, hyperthermia, and athletes: science and practice. *Journal of Athletic Training*. 1996;31(3):248–252.

Nakamura K, Okada A, Watanabe H, et al. In-hospital mortality of heat-related disease associated with wet bulb globe temperature: a Japanese nationwide inpatient data analysis. *Int J Biometeorol*. 2025;69(4):873–884. doi:10.1007/s00484-025-02867-x.

Nash D, Hughes MG, Butcher L, et al. IL-6 signaling in acute exercise and chronic training: Potential consequences for health and athletic performance. *Scand J Med Sci Sports*. 2023;33(1):4–19. doi:10.1111/sms.14241.

Naulleau C, Jeker D, Pancrate T, et al. Effect of Pre-Exercise Caffeine Intake on Endurance Performance and Core Temperature Regulation During Exercise in the Heat: A Systematic Review with Meta-Analysis. *Sports Med*. 2022;52(10):2431–2445. doi:10.1007/s40279-022-01692-1.

Naumann J, Kruza I, Denkel L, Kienle G, Huber R. Effects and feasibility of hyperthermic baths in comparison to exercise as add-on treatment to usual care in depression: a randomised, controlled pilot study. *BMC Psychiatry*. 2020;20(1):536. doi:10.1186/s12888-020-02941-1.

Naumann J, Kruza I, Denkel L, Kienle G, Huber R. Effects and feasibility of hyperthermic baths in comparison to exercise as add-on treatment to usual care in depression: a randomised, controlled pilot study. *BMC Psychiatry*. 2020;20(1):536. doi:10.1186/s12888-020-02941-1.

NCAA Sport Science Institute. Preventing Catastrophic Injury and Death in Collegiate Athletes. 2019. https://ncaaorg.s3.amazonaws.com/ssi/injury_prev/SSI_PreventingCatastrophicInjuryBooklet.pdf.

New York Times. Bronx planting caps off a drive to add a million trees. *The New York Times.* https://www.nytimes.com/2015/10/21/nyregion/new-york-city-prepares-to-plant-one-millionth-tree-fulfilling-a-promise.html?ref=nyregion.

Newman RW. Why man is such a sweaty and thirsty naked animal: a speculative review. *Hum Biol.* 1970;42(1):12–27..

News ABC. Jury Rules Against Radio Station After Water-Drinking Contest Kills Calif. Mom. ABC News.

NHS Business Services Authority. NHS Releases 2023/2024 mental health medicines statistics for England. NHS. Accessed July 17, 2025. https://media.nhsbsa.nhs.uk/press-releases/5171d616-95ea-4282-959b-15f8bfed6a0f/nhs-releases-2023-24-mental-health-medicines-statistics-for-england.

Noakes T. Waterlogged: the serious problem of overhydration in endurance sports. *Human Kinetics*; 2012:428.

Noakes TD. Hydration in the marathon: using thirst to gauge safe fluid replacement. *Sports Med.* 2007;37(4–5):463–466. doi:10.2165/00007256-200737040-00050.

Noakes TD, Goodwin N, Rayner BL, Branken T, Taylor RK. Water intoxication: a possible complication during endurance exercise. *Med Sci Sports Exerc.* 1985;17(3):370–5.

Nolte HW, Noakes TD, Van Vuuren B. Trained humans can exercise safely in extreme dry heat when drinking water ad libitum. *J Sports Sci.* 2011;29(12):1233–1241. doi:10.1080/02640414.2011.587195.

Nolte HW, Nolte K, Hew-Butler T. Ad libitum water consumption prevents exercise-associated hyponatremia and protects against dehydration in soldiers performing a 40-km route-march. *Mil Med Res.* 2019;6(1):1. doi:10.1186/s40779-019-0192-y

Nyer MB, Hopkins LB, Nagaswami M, et al. A Randomized Controlled Trial of Community-Delivered Heated Hatha Yoga for Moderate-to-Severe Depression. *J Clin Psychiatry.* 2023;84(6):22m14621. doi:10.4088/JCP.22m14621.

O'Donnell M. Scientologist's Treatments Lure Firefighters. *New York Times.* 2003/10/04/:A1. Accessed March 27, 2025. https://www.nytimes.com/2003/10/04/nyregion/scientologist-s-treatments-lure-firefighters.html.

O'Kelly E. SAUNA: The Power of Deep Heat. *Welbeck Balance*; 2023.

O'Connor FG, DeGroot DW. Heat-Related Illness in Athletes. *JAMA.* 2024;332(8):664–665. doi:10.1001/jama.2024.9991.

Oliver DM, McDougall CW, Robertson T, Grant B, Hanley N, Quilliam RS. Self-reported benefits and risks of open water swimming to health, wellbeing and the environment: Cross-sectional evidence from a survey of Scottish swimmers. *PLoS ONE.* 2023;18(8):e0290834. doi:10.1371/journal.pone.0290834.

Patrick R. Dr. Jari Laukkanen on Sauna Use for the Prevention of Cardiovascular & Alzheimer's Disease. Accessed May 20, 2023. https://www.foundmyfitness.com/episodes/jari-laukkanen.

Patrick RP, Johnson TL. Sauna use as a lifestyle practice to extend healthspan. *Experimental Gerontology.* 2021;154:111509. doi:10.1016/j.exger.2021.111509.

Périard JD, DeGroot D, Jay O. Exertional heat stroke in sport and the military: epidemiology and mitigation. *Experimental Physiology.* 2022;107(10):1111–1121. doi:10.1113/EP090686.

Périard JD, Eijsvogels TMH, Daanen HAM. Exercise under heat stress: thermoregulation, hydration, performance implications, and mitigation strategies. *Physiological Reviews*. 2021;101(4):1873–1979. doi:10.1152/physrev.00038.2020.

Périard JD, Racinais S, Sawka MN. Adaptations and mechanisms of human heat acclimation: Applications for competitive athletes and sports. *Scandinavian Journal of Medicine & Science in Sports*. 2015;25(S1):20–38. doi:10.1111/sms.12408.

Périard JD, Travers GJS, Racinais S, Sawka MN. Cardiovascular adaptations supporting human exercise-heat acclimation. *Autonomic Neuroscience*. 2016;196:52–62. doi:10.1016/j.autneu.2016.02.002.

Perrotta AS, White MD, Koehle MS, Taunton JE, Warburton DER. Efficacy of Hot Yoga as a Heat Stress Technique for Enhancing Plasma Volume and Cardiovascular Performance in Elite Female Field Hockey Players. *J Strength Cond Res*. 2018;32(10):2878–2887. doi:10.1519/JSC.0000000000002705.

Peterson C. Meet the animals that can handle extreme heat. *National Geographic*. 2023.

Phillips PA, Rolls BJ, Ledingham JGG, Morton JJ. Body fluid changes, thirst and drinking in man during free access to water. *Physiology & Behavior*. 1984;33(3):357–363. doi:10.1016/0031-9384(84)90154-9.

Piil JF, Mikkelsen CJ, Junge N, Morris NB, Nybo L. Heat Acclimation Does Not Protect Trained Males from Hyperthermia-Induced Impairments in Complex Task Performance. *Int J Environ Res Public Health*. 2019;16(5):716. doi:10.3390/ijerph16050716.

Price L. Sweat House, Co. Wicklow. *Journal of the Royal Society of Antiquaries of Ireland*. 1952;82(2):180–181.

Racinais S, Moussay S, Nichols D, et al. Core temperature up to 41.5°C during the UCI Road Cycling World Championships in the heat. *Br J Sports Med*. 2019;53(7):426–429. doi:10.1136/bjsports-2018-099881.

Rae DE, Knobel GJ, Mann T, Swart J, Tucker R, Noakes TD. Heatstroke during Endurance Exercise: Is There Evidence for Excessive Endothermy? *Medicine & Science in Sports & Exercise*. 2008;40(7):1193–1204. doi:10.1249/MSS.0b013e31816a7155.

Rastogi S, Singh RH. Principle of Hot (Ushna) and Cold (Sheeta) and Its Clinical Application in Ayurvedic Medicine. In: Yavari M, ed. *Hot and Cold Theory: The Path Towards Personalized Medicine*. Springer International Publishing; 2021:39–55.

Ritchie H. Air conditioning causes around 3% of greenhouse gas emissions. How will this change in the future? *Our World in Data*. Accessed July 22, 2025. https://ourworldindata.org/air-conditioning-causes-around-greenhouse-gas-emissions-will-change-future.

Ritossa F. A New Puffing Pattern Induced by Temperature Shock and DNP in Drosophila. Experientia. 1962:571–573. doi:https://doi.org/10.1007/BF02172188.

Roberts JW, Powlovich L, Sheybani N, LeBlang S. Focused ultrasound for the treatment of glioblastoma. *J Neurooncol*. 2022;157(2):237–247. doi:10.1007/s11060-022-03974-0.

Roberts LA, Raastad T, Markworth JF, et al. Post-exercise cold water immersion attenuates acute anabolic signalling and long-term adaptations in muscle to strength training. *The Journal of Physiology*. 2015;593(18):4285–4301. doi:10.1113/JP270570.

Rollins H. The Rollins Column: August Ode. *LA Weekly*. 2011.

Romanovsky AA. The thermoregulation system and how it works. *Handb Clin Neurol.* 2018;156:3–43. doi:10.1016/B978-0-444-63912-7.00001-1.

Rønnestad BR, Hamarsland H, Hansen J, et al. Five weeks of heat training increases haemoglobin mass in elite cyclists. *Experimental Physiology.* 2021;106(1):316–327. doi:10.1113/EP088544.

Rønnestad BR, Lid OM, Hansen J, et al. Heat suit training increases hemoglobin mass in elite cross-country skiers. *Scandinavian Journal of Medicine & Science in Sports.* 2022;32(7):1089–1098. doi:10.1111/sms.14156.

Rønnestad BR, Urianstad T, Hamarsland H, et al. Heat Training Efficiently Increases and Maintains Hemoglobin Mass and Temperate Endurance Performance in Elite Cyclists. *Med Sci Sports Exerc.* 2022;54(9):1515–1526. doi:10.1249/MSS.0000000000002928.

Ruberg S. Three Hikers Die in Grand Canyon National Park in Less Than a Month. *The New York Times.* Accessed June 27, 2025. https://www.nytimes.com/2024/07/09/us/grand-canyon-hiker-death.html.

Sambon LW. Acclimatization of Europeans in Tropical Lands. *The Geographical Journal.* 1898;12(6):589–599.

Sarris J, De Manincor M, Hargraves F, Tsonis J. Harnessing the Four Elements for Mental Health. *Front Psychiatry.* 2019;10:256. doi:10.3389/fpsyt.2019.00256.

Schaffer A. The Scientologists' dubious detox program. *Slate.* https://slate.com/technology/2004/10/the-scientologists-dubious-detox-program.html.

Schnare DW, Denk G, Shields M, Brunton S. Evaluation of a detoxification regimen for fat stored xenobiotics. *Medical Hypotheses.* 1982;9(3):265–282. doi:10.1016/0306-9877(82)90156-6.

Schneider SM. Heat acclimation: Gold mines and genes. *Temperature (Austin).* 2016;3(4):527–538. doi:10.1080/23328940.2016.1240749.

Schroeder R. Crystal Water Bottles Are A Wellness Trend We Can Get Behind. *Elle.* May 30, 2022. https://www.elle.com/uk/beauty/body-and-physical-health/a28242635/crystal-water-bottle-wellness-trend.

Shakespeare J. Bend it like Bikram. *The Guardian.* June 11, 2006. https://www.theguardian.com/lifeandstyle/2006/jun/11/healthandwellbeing.features1.

Shattock MJ, Tipton MJ. 'Autonomic conflict': a different way to die during cold water immersion? *The Journal of Physiology.* 2012;590(14):3219–3230. doi:10.1113/jphysiol.2012.229864.

Sherwood SC, Huber M. An adaptability limit to climate change due to heat stress. *Proc Natl Acad Sci U S A.* 2010;107(21):9552–9555. doi:10.1073/pnas.0913352107.

Singleton M. *Yoga Body: The Origins of Modern Posture Practice.* Oxford University Press; 2010.

Smart R. Ditch the tie and reduce the AC—Japan's Cool Biz gets summer hell just about right. *Quartz.* Accessed July 22, 2025. https://qz.com/465327/ditch-the-tie-and-reduce-the-ac-japans-cool-biz-gets-summer-hell-just-about-right?utm_source=chatgpt.com.

Souetre E, Salvati E, Wehr TA, Sack DA, Krebs B, Darcourt G. Twenty-four-hour profiles of body temperature and plasma TSH in bipolar patients during depression and during remission and in normal control subjects. *Am J Psychiatry.* 1988;145(9):1133–1137. doi:10.1176/ajp.145.9.1133.

Souetre E, Salvati E, Wehr TA, Sack DA, Krebs B, Darcourt G. Twenty-four-hour profiles of body temperature and plasma TSH in bipolar patients during depression and during remission and in normal control subjects. *Am J Psychiatry*. 1988;145(9):1133–1137. doi:10.1176/ajp.145.9.1133.

Speedy DB, Noakes T. Hyponatremia in ultradistance triathletes. *Medicine & Science in Sports & Exercise*. 1999;31(6):809–815.

Speedy DB, Noakes TD, Schneider C. Exercise-associated hyponatremia: A review. *Emergency Medicine*. 2001;13(1):17–27. doi:10.1046/j.1442-2026.2001.00173.x.

Speedy DB, Rogers IR, Noakes TD, et al. Exercise-induced hyponatremia in ultradistance triathletes is caused by inappropriate fluid retention. *Clin J Sport Med*. 2000;10(4):272–278. doi:10.1097/00042752-200010000-00009.

Spelman College. UrbanHeatATL. Accessed July 11, 2025. https://urbanheatatl.org.

Srámek P, Simecková M, Janský L, Savlíková J, Vybíral S. Human physiological responses to immersion into water of different temperatures. *Eur J Appl Physiol*. 2000;81(5):436–442. doi:10.1007/s004210050065.

Staff R. Football player from Georgia dies at camp. *Ocala Star Banner*. 2011. https://www.ocala.com/story/news/local/2011/08/03/football-player-from-georgia-dies-at-camp/64287178007/.

Staff W. Mom dies from drinking too much water, family says. 2023. https://wwwfox5 vegascom.

Stapleton JM, Poirier MP, Flouris AD, et al. Aging impairs heat loss, but when does it matter? *J Appl Physiol* (1985). 2015;118(3):299–309. doi:10.1152/japplphysiol.00722.2014.

Steadman RG. The Assessment of Sultriness. Part I: A Temperature-Humidity Index Based on Human Physiology and Clothing Science. J Appl Meteor. 1979;18(7):861–873. doi:10.1175/1520-0450(1979)018<0861:TAOSPI>2.0.CO;2.

Stocks JM, Taylor NA, Tipton MJ, Greenleaf JE. Human physiological responses to cold exposure. Aviat Space Environ Med. 2004;75(5):444–57.

Stouhi D. What is a Traditional Windcatcher? *Arch Daily* blog. 2021. https://www.archdaily.com/971216/what-is-a-traditional-windcatcher.

Syman S. *The Subtle Body: The Story of Yoga in America*. 1st ed. Farrar, Straus and Giroux; 2010:390.

Szuba MP, Guze BH, Baxter LR. Electroconvulsive therapy increases circadian amplitude and lowers core body temperature in depressed subjects. *Biol Psychiatry*. 1997;42(12):1130–1137. doi:10.1016/s0006-3223(97)00046-2.

Tankersley CG, Smolander J, Kenney WL, Fortney SM. Sweating and skin blood flow during exercise: effects of age and maximal oxygen uptake. *J Appl Physiol* (1985). 1991;71(1):236–42. doi:10.1152/jappl.1991.71.1.236.

Tao L, Su Y, Chen X, Tian F. Processes, spatial patterns, and impacts of the 1743 extreme-heat event in northern China: from the perspective of historical documents. *Clim Past*. 2024;20(11):2455–2471. doi:10.5194/cp-20-2455-2024.

Tarrant K, Southern K. Wim Hof started a cold water therapy trend. Our daughters died trying it. *The Sunday Times*. 2024/06/23/. Accessed April 23, 2025. https://www.thetimes.com/uk/society/article/wim-hof-iceman-breathing-method-cold-water-therapy-killed-9tnvc5w8q.

Taylor A, O'Malley M, O'Callaghan R, Goodwin J. Exploring the Use of Sea Swimming as an Intervention With Young People With Mental Health Challenges: A Qualitative Descriptive Study. *Int J Ment Health Nurs.* 2025;34(1):e70000. doi:10.1111/inm.70000.

Thompson A. This Hot Summer Is One of the Coolest of the Rest of Our Lives. *Scientific American.* 2022.

Tipton MJ. The Initial Responses to Cold-Water Immersion in Man. *Clinical Science.* 1989;77(6):581–588. doi:10.1042/cs0770581.

Tipton MJ, Collier N, Massey H, Corbett J, Harper M. Cold water immersion: kill or cure? *Experimental Physiology.* 2017;102(11):1335–1355. doi:10.1113/EP086283

Tucker R, Rauch L, Harley YX, Noakes TD. Impaired exercise performance in the heat is associated with an anticipatory reduction in skeletal muscle recruitment. *Pflugers Arch.* J2004;448(4):422–30. doi:10.1007/s00424-004-1267-4.

Tuomilehto, J., Sarti, C., Narva, E. V., Salmi, K., Sivenius, J., Kaarsalo, E., Salomaa, V., & Torppa, J. (1992). The FINMONICA Stroke Register. Community-based stroke registration and analysis of stroke incidence in Finland, 1983–1985. *American journal of epidemiology, 135(11),* 1259–1270. https://doi.org/10.1093/oxfordjournals.aje.a116232.

Ukai T, Iso H, Yamagishi K, et al. Habitual tub bathing and risks of incident coronary heart disease and stroke. *Heart.* 2020;106(10):732–737. doi:10.1136/heartjnl-2019-315752.

United States Environmental Protection Agency. Climate Change Indicators: Heat Waves. https://www.epa.gov/climate-indicators/climate-change-indicators-heat-waves

University of Eastern Findland. Kuopio Ischaemic Heart Disease Risk Factor (KIHD) Study, *Nutrition.* University of Eastern Finland. Accessed July 10, 2025. https://uefconnect.uef.fi/en/kuopio-ischaemic-heart-disease-risk-factor-kihd-study-nutrition/#publications.

Uranga F, Douglas E. As Texas swelters, local rules requiring water breaks for construction workers will soon be nullified. *The Texas Tribune.* 2023. https://www.texastribune.org/2023/06/16/texas-heat-wave-water-break-construction-workers/.

US Department of Health and Human Services. Heat.gov. Accessed July 8, 2024. https://www.heat.gov.

US Energy Information Administration. Nearly 90% of U.S. households used air conditioning in 2020. 2022/05/31. https://www.eia.gov/todayinenergy/detail.php?id=52558.

Vahtla A. Finns' Friday night sauna in part behind day's peak power prices in Estonia. ERR News. 2024. https://news.err.ee/1609213489/finns-friday-night-sauna-in-part-behind-day-s-peak-power-prices-in-estonia.

Valtin H. "Drink at least eight glasses of water a day." Really? Is there scientific evidence for "8 x 8"? *Am J Physiol Regul Integr Comp Physiol.* 2002;283(5):R993–1004. doi:10.1152/ajpregu.00365.2002.

Van Der Zee J. Heating the patient: a promising approach? *Annals of Oncology.* 2002;13(8):1173–1184. doi:10.1093/annonc/mdf280.

Van Lanen, J. M. 2017. Ski hunters of Siberia: Self-reliance in Central Asia's Altai Mountains. *Alaska Fish & Wildlife News.* February 2017. Alaska Department of Fish and Game, Division of Wildlife Conservation, Juneau. http://www.adfg.alaska.gov/index.cfm?adfg=wildlifenews.view_article&articles_id=810.

Wall BA, Watson G, Peiffer JJ, Abbiss CR, Siegel R, Laursen PB. Current hydration guidelines are erroneous: dehydration does not impair exercise performance in the heat. *Br J Sports Med.* 2015;49(16):1077–1083. doi:10.1136/bjsports-2013-092417.

Wallace DC. A mitochondrial paradigm of metabolic and degenerative diseases, aging, and cancer: a dawn for evolutionary medicine. *Annu Rev Genet.* 2005;39:359–407. doi:10.1146/annurev.genet.39.110304.095751.

Wallace-Wells D. *The Uninhabitable Earth: Life After Warming.* New York. 2017.

Wallace-Wells D. The Mysteriously Low Death Toll of the Heat Waves in India and Pakistan. *New York Times.* 2022.

Ward NG, Doerr HO, Storrie MC. Skin conductance: a potentially sensitive test for depression. *Psychiatry Res.* 1983;10(4):295–302. doi:10.1016/0165-1781(83)90076-8.

Watson P, Hasegawa H, Roelands B, Piacentini MF, Looverie R, Meeusen R. Acute dopamine/noradrenaline reuptake inhibition enhances human exercise performance in warm, but not temperate conditions. *J Physiol.* 2005;565(Pt 3):873–83. doi:10.1113/jphysiol.2004.079202.

Watson P, Whale A, Mears SA, Reyner LA, Maughan RJ. Mild hypohydration increases the frequency of driver errors during a prolonged, monotonous driving task. *Physiology & Behavior.* 2015;147:313–318. doi:10.1016/j.physbeh.2015.04.028.

WCBV 5 Boston. Doctors: Marathoner Died From Too Much Water. Hyponatremia a Danger in Long Distance Sports. https://www.wcvb.com/article/doctors-marathoner-died-from-too-much-water/8126646.

Weill Cornell Medical College. The Rise and Decline of Psychiatric Hydrotherapy (Page 3). http://psych-history.weill.cornell.edu/osk_die_lib/hydrotherapy/Page3.html.

Wellbeing Research Centre at the University of Oxford. World Happiness Report: Data Sharing. Accessed July 16, 2025. https://www.worldhappiness.report/data-sharing/.

Wharam PC, Speedy DB, Noakes TD, Thompson JMD, Reid SA, Holtzhausen L-M. NSAID use increases the risk of developing hyponatremia during an Ironman triathlon. *Med Sci Sports Exerc.* 22006;38(4):618–622. doi:10.1249/01.mss.0000210209.40694.09.

Wheeler PE. The thermoregulatory advantages of hominid bipedalism in open equatorial environments: the contribution of increased convective heat loss and cutaneous evaporative cooling. *Journal of Human Evolution.* 1991;21(2):107–115. doi:10.1016/0047-2484(91)90002-D.

White J. Standing Tall: Swoboda Overcomes Life-Threatening Ordeal. *University of Virginia Sports.* 2024. https://virginiasports.com/news/2019/08/23/jeff-white-swoboda/.

White RH, Anderson S, Booth JF, et al. The unprecedented Pacific Northwest heatwave of June 2021. *Nat Commun.* 2023;14(1):727. doi:10.1038/s41467-023-36289-3.

Whitfield AHN. Too much of a good thing? The danger of water intoxication in endurance sports. *Br J Gen Pract.* 2006;56(528):542–545.

Whitford WG. Sweating responses in the chimpanzee (Pan troglodytes). Comparative Biochemistry and Physiology Part A: Physiology. 1976;53(4):333–336. doi:10.1016/S0300-9629(76)80151-X.

Wijnen AHC, Steennis J, Catoire M, Wardenaar FC, Mensink M. Post-Exercise Rehydration: Effect of Consumption of Beer with Varying Alcohol Content on Fluid Balance after Mild Dehydration. *Front Nutr.* 2016;3:45. doi:10.3389/fnut.2016.00045.

Wilke K, Martin A, Terstegen L, Biel SS. A short history of sweat gland biology. *Intern J of Cosmetic Sci*. 2007;29(3):169–179. doi:10.1111/j.1467-2494.2007.00387.x.

Williams DO. The Paradise Paradox: A Colorado ski town's struggle with suicide, depression. *Colorado Springs Gazette*. Accessed Dec 2, 2024. https://gazette.com/health/the-paradise-paradox-a-colorado-ski-towns-struggle-with-suicide-depression/article_5d095e24-1138-11ea-936d-83ac14c08352.html.

Williams LE, Bargh JA. Experiencing Physical Warmth Promotes Interpersonal Warmth. *Science*. 2008;322(5901):606–607. doi:10.1126/science.1162548.

Wright FL. *The Natural House*. Horizon Press, New York. 1954.

WSB-TV 2 Atlanta. Student Athlete Dies of Heat Stroke. Atlanta, GA: WSB-TV; 2011.

Wyndham CH. Effect of acclimatization on the sweat rate-rectal temperature relationship. *J Appl Physiol*. 1967;22(1):27–30. doi:10.1152/jappl.1967.22.1.27.

Wyndham CH. Heat stroke and hyperthermia in marathon runners. *Ann N Y Acad Sci*. 1977;301:128–138. doi:10.1111/j.1749-6632.1977.tb38192.x.

Wyndham CH, Strydom NB. The danger of an inadequate water intake during marathon running. *South African Medical Journal*. 1969;43(29):893–896. doi:10.10520/AJA20785135_33411.

Wynne JL, Wilson PB. Got Beer? A Systematic Review of Beer and Exercise. *International Journal of Sport Nutrition and Exercise Metabolism*. 2021;31(5):438–450. doi:10.1123/ijsnem.2021-0064.

Yamada Y, Zhang X, Henderson MET, et al. Variation in human water turnover associated with environmental and lifestyle factors. *Science*. 2022;378(6622):909–915. doi:10.1126/science.abm8668.

Yang S, Liu J, Gu Z, Liu P, Lan Q. Physiological and Metabolic Adaptation to Heat Stress at Different Altitudes in Yaks. *Metabolites*. 2022;12(11):1082. doi:10.3390/metabo12111082.

Yegül FK. *Bathing in the Roman World*. 1. publ ed. Cambridge Univ. Press; 2010:256.

Zech M, Bosel S, Tuthorn M, et al. Sauna, sweat and science—quantifying the proportion of condensation water versus sweat using a stable water isotope ($(2)H/(1)H$ and $(18)O/(16)O$) tracer experiment. *Isotopes Environ Health Stud*. 2015;51(3):439–447. doi:10.1080/10256016.2015.1057136.

Zhao Q, Guo Y, Ye T, et al. Global, regional, and national burden of mortality associated with non-optimal ambient temperatures from 2000 to 2019: a three-stage modelling study. *Lancet Planet Health*. 2021;5(7):e415-e425. doi:10.1016/S2542-5196(21)00081-4.

Zschaeck S, Beck M. Whole-Body Hyperthermia in Oncology: Renaissance in the Immunotherapy Era? In: Vaupel P, ed. *Water-filtered Infrared A (wIRA) Irradiation*. Springer International Publishing; 2022:107–115.

INDEX

NOTE: Page references in *italics* refer to figures and photos. Page references including "n" refer to footnotes at bottom of pages.

Aaland, Mikkel, xviii–xix, 165–166, 233
Abbott, Greg, 133
acclimation. *see* heat adaptation
"Acclimatization of Europeans in Tropical Lands," 22
Acerbi, Giuseppe, *163*
age and aging
   heat adaptation and, 140, 149–151
   heat and effect on, 9, 137, 228
   hydration and, 96
air-conditioning, 57–68
   alternatives to, 65–66
   climate adaptation prior to, 57–58, 61–62
   climate control with, 62–64
   early ice inventions, 59–61, *60*
   heat adaptation and, 147
   heat-island effect and, 64–66
   indoor lifestyle due to, 66–68
   invention of, 61
Albertson, Clayton ("CJ"), xxi, 71–73, 85, 86

Alzheimer's disease (dementia), 45, 49–50, 53, 54
America (Ginsberg), 228
American College of Sports Medicine (ACSM), 94, 98, 102
Anderson, S. A., 125n
antidepressants, 105, 206n, 210–212, 216–217
antiperspirant, 67
apocrine sweat glands, 26–27
Archimedes Banya, 219
Arctic peoples, 142–143
arteries, 51
Attia, Peter, xviii, 109n
Auerbach, Glenn, 198–200
Australopithecus, 35

Bain, Anthony, 45, 149, 151
bain de surprise, 191, 199, 221
Bangkok, Thailand, 145–146, *146*
Banks, Joseph, 21
basketball, heatstroke from, 130n

INDEX

Baths of Caracalla, 164
Baths of Diocletian, 164
Battle Creek Sanitarium, 191
Beck's Depression Inventory, 222, 227
Bedford, Sylvia, 76–77
beer, 110
behavioral thermoregulation, 25
Benson, Herbert, 213–214
Bent, Robbie, 233–234
Bent, Shannon, 233
Berlin Marathon, 100
Best, Andrew, 27, 32
beta-endorphin, 52
beverage industry. *see* hydration; *individual names of brands*
bicycling. *see* cycling
Biden, Joe, 132
Bikram yoga, 175–179, 180, 211–212, 224n
Blagden, Charles, 20, 21
blood. *see also* heart health
  blood plasma, 84–86, 183
  blood pressure, 51–52, 88–89, 123
  blood salinity, 94
  circulation, 27–28, 50–51, 73–74, 194
  hemoglobin mass, 86–87
Blood, Urine, and Sweat Study (BUS), 184–185
Bloomberg, Michael, 66
"Bluetits" swimmers, 197
Blummenfelt, Kristian, 79
Bode (dog), 26–27, 31, 34, 81, 149
body positivity, 159
body temperature and measurement. *see also* sweat; thermoregulation
  average human body temperature, 28n
  cold plunging and core temperature, 196
  CORE (for heat flux measurement), 82–84
  core temperature, 10–12, 16, 196
  dehydration and, 98
  depression and, 204–206, *205*
  fever, benefits of, 215–218, 221
  monitoring, for sports practice, 129–130
  rectal probe for, 10–12, 80–81
  resting body temperature, 118
  skin temperature, 28, 82, 85, 198
  temperature-sensing capsule, 10, 16
  Tummo studied with, 214
  uncompensable conditions of, 9–10, 116
body weight, 96, 99–101, 108, 148n
bordellos, 166
Born to Run (McDougall), 36
Boston Marathon, 71–73, 93–95
Botswana (Kalahari Desert), 36n
boxing, 78n
brain. *see also* mental health; nervous system
  as Central Governor, 76–78, 85, 101
  clinical hyponatremia, 105
  cytokines, 215
  dorsal raphe nucleus, 215
  exercise-associated hyponatremic encephalopathy (EAHE), 94–95
  serotonin, 216
Brazil, *146*, 147
Breakthrough Institute, 65, 147
breathwork, Tummo, 213–214
British Medical Journal, 43
Brown, Patrick, 65, 147
brown fat, 193–194
Bu, Olav Aleksander, 79, 83, 86
Business Slalom, 162, 234
Butch (Turtle Boy), 116
Byculla Club, 59

Cade, Robert, 97
caffeinated beverages, 109
California, heat-safety guidelines of, 132
CamelBak, 114–115
camels, 26
Caracalla (emperor of Rome), 164
cardiovascular (CV) drift, 13. *see also* heart health
Carney, Scott, 192
Carrier, David, 35–36
Carrier, Willis, 61

Carrier Corporation, 61
Casa, Douglas, 15, 74, 102–103, 127, 134, 193
Cass, Meridith, 108–109
Cedar + Stone, 187–188
Celsius, Anders, 23
Central Governor, 76–78, 85, 101
Cheuvront, Stephen, 107
CHILL'D (Cold and Heat to Investigate Lowered Levels of Depression) study, 203–204, 206–207, 211, *216*, 216–219, 222–225, 227–228
chimpanzees, 28, 32
China, 133, 147
Choudhury, Bikram, 175–176, 177, 178
Church of Scientology, 181–182, 185
Clearlight Dome infrared sauna, 219
climate change
  air-conditioning and effect on, 63
  heat acclimation and, 145
  heat-island effect vs., 65
  heat stress and, 139
  heat waves (Europe, 2023), 140
  source of, 6
Coca-Cola, 104, 109
Cold and Heat to Investigate Lowered Levels of Depression (CHILL'D) study, 203–204, 206–207, *216*, 216–219, 222–225, 227–228
cold plunging, 187–202
  in American sauna culture, 187–189, 198–202
  brown fat and, 193–194
  cold shock response, 189–190
  cold swimming and, for depression, 208
  contrast therapy and, 194–195
  core temperature and, 196
  in Finnish sauna culture, 42, 155
  for heat stroke, 75, 127, 193
  history of cold-water bathing, 191
  for inflammation, 195
  large-scale study (Netherlands) of, 197–198
  mental health benefits of, 196, 198
  for non-athletes, 197
  social benefits of, 199–200
  Wim Hof Method of, 191–194, 196
cold-related mortality, geographical distribution, 143–149, *146*
ColdVest, 134
Coley, William, 218
Colorado, heat-safety guidelines of, 132
Comrades ultramarathon, 95
"Cool Biz" initiative, 66
Cooper, Earl "Bud," 124, 127, 129
CORE device, 82–83, 88
core temperature, 10–12, 16, 196. *see also* body temperature and measurement
"corn sweat," 64n
Covid, 45, 71, 73, 79, 171, 197–198, 204, 233
C-reactive protein (CRP), 51
critical temperature limit, 24
Cullen, William, 60
cycling. *see also* heat adaptation; Hotter'N Hell Hundred bicycle marathon
  Falmouth Road Race, 74
  internal body temperature of cyclists, 24
  Tour de France, 77, 78, 83, 96, 109, 191
cytokines, 215

Darwin, Charles, 143
David-Néel, Alexandra, 213, 213n
default mode network, 221
DeGroot, David, 100
dehydration symptoms, 100–101, 160. *see also* hydration
dementia (Alzheimer's disease), 45, 49–50, 53, 54
Denisovans, 34, 230n
deodorant, 67
Department of Defense, US, 8
depression. *see* mental health
DeSantis, Ron, 133

detoxification, 175–185
  Bikram yoga and, 175–179, 180
  Blood, Urine, and Sweat Study (BUS), 184–185
  Church of Scientology on, 181–182, 185
  Galen on sweating as excretory function, 182–183
  HOTWORX and, 179–181
  "sweating out toxins" claims, 181–185
Ditropan, 206
dogs, 26–27, 31, 34, 81, 149
dorsal raphe nucleus, 215
Dreosti, Aldo, 12–13
dynorphin, 52

Eagle County, Colorado, 209
Eastern medicine practices, 218
eccrine sweat glands, 27–34, *30*, 183
ectotherms, 23
Egypt (ancient), 62, 191
Eidson, Michael, 114
electricity, air-conditioning and, 62–63
Ellis, Henry, 20–21, 22, 59
Elomaa, Risto, 39–42, 158
endothelium, 51, 88–89
endotherms, 23–24
endurance, 69–151
  heat adaptation, 137–151 (*see also* heat adaptation)
  heat illness issues of, 121–135 (*see also* heat illness and heatstroke)
  heat training for, 71–89 (*see also* heat adaptation)
  in Hotter'N Hell Hundred bicycle marathon, 111–119 (*see also* Hotter'N Hell Hundred bicycle marathon)
  hydration and, 91–110 (*see also* hydration)
England, 145–146, *146*
Engrailed-1 (*EN1*) gene, 33
environmental physiology, 45
ephedra, 148n
Europe, heat waves (2023), 140
European Hydration Institute, 104

European Union, 133
evaporation of sweat, 29–32, *30*, 74. *see also* sweat
Everts, Sarah, 184–185
evolution
  climate adaptation by humans and, 57–58
  DNA study of sweat, 230–231
  heat adaptation and, 141–143
  of human sweating, 32–37
  mammals and different methods of body cooling, 25–27, 31, 34
excess deaths, defined, 143
exercise. *see also individual types of sports*
  exercise-associated hyponatremic encephalopathy (EAHE), 94–95
  for heat adaptation, 150–151
  heat adaptation vs. aerobic fitness, 76–77, 81
  heated exercise and calorie burning, 180n
  inflammation relief and, 215
  studying effects of, xviii, 46
  warming up for, 195
exertional heatstroke, 15n. *see also* heat illness and heatstroke

Fahrenheit, Daniel Gabriel, 22–23
Falmouth (Massachusetts) Road Race, 74–75
fever, 19, 215–218, 221
FIFA World Cup, 133
fight-or-flight response, 29. *see also* nervous system
Finland
  birth rate in, 165
  life expectancy in, 43
  World Happiness Report and, 160
Finland, sauna in
  body positivity and, 159
  cold plunge routines, 42, 155
  as cultural practice, xi–xiv, 39–43, 157, 160–161
  Finnish sauna vs. other sauna traditions, 54–55

history of Roman baths and, *163*,
  163–167
Löyly (Helsinki Harbor), 157–
  160, 167
löyly (steam), xii–xiv, 40–41, 50, 158,
  168, 169–170
  at Panorama Landscape Hotel,
    161–162
  Rajaportti (Tampere), xi–xiv, xiii, 157,
    162–163, 167–173, *168, 169*
  research on, 43–51, 207
  "rounds" of, 41, 156
  savusauna (smoke sauna), 158
  sisu (grit) and, 41, 172, 216
  social benefits of, 156–157, 171–173
Finnish Sauna Society (*Suomalainen
  Saunaseura*), 39–42, 158, 191
firefighters, detoxification by, 185
Fisher, Kathleen, 140, 149
Fitzgerald (Georgia) High School
  football (Purple Hurricanes),
  121–124
Florida, 133
football, 121–130, 137–138
Fordyce, George, 19–20, 21, 23
Foster, Craig, 36n
Foster, H. (British Army doctor), 43
Foster, Simmie, 53
Four Seasons Hotel (Minneapolis), 188
France
  air-conditioning in, 63–64
  Tour de France, 77, 78, 83, 96,
    109, 191
fruit flies, 53n, 54

Galen, 182–183
Galileo, 22
#GallonChallenge, 93
Gatorade, 93, 94, 97–98, 109
Gebreselassie, Haile, 100
genetics
  DNA/prehistoric jewelry study
    (Netherlands), 230
  mitochondria and heat adaptation,
    142–143
  sweating and, 32–34

Genuis, Stephen, 184
geographical distribution, temperature
  and mortality, 143–149, *146*
Georgia
  heat measured in (colonial era),
    20–21, 59
  heat-safety guidelines in, 126–130
  trees of Atlanta, 64
Georgia High School Association, 129
German sauna, 54
Gifford, Bill. *see also* heat adaptation
  heat adaptation by, 117–119
  Outlive, xviii, 109n
  photos, *118, 119, 216*
Ginsberg, Allen, 228
gold miners, 12–13
Gorrie, John, 59–61
The Great Dance documentary
  (Foster), 36n
grit (*sisu*), 41, 172, 216
Grundstein, Andrew, 124–125, 129

hair, sweat and, 26, 32–35
hammams (baths of Muslim world),
  165–166
Harvard University, 53, 211–214
health benefits of heat, 39–56
  Alzheimer's (dementia) and, 45, 49–
    50, 53, 54
  body's response to sauna, 50–52
  cold plunging and, 187–202 (*see also*
    cold plunging)
  for depression, 203–225 (*see also*
    mental health)
  detoxification and, 175–185 (*see also*
    detoxification)
  Finnish sauna tradition and, 39–43,
    171–173
  Finnish sauna vs. other sauna
    traditions for, 54–55
  heart disease and, 42–45, 47–52, 54
  heat shock proteins, 52–54, 80–81
  heat therapy concept and, 45
  hot baths and hot tub use for, 55–56
  KIHD research study and follow-up,
    43–51

health benefits of heat (*cont.*)
   mental health benefits, 203–225 (*see also* mental health)
heart health
   cold water and, 192–193
   dehydration and heart rate, 98
   heat and benefits to, 42–45, 47–52, 54
   heat training and, 89
   nervous system and, 207
heat, power of. *see* power of heat
heat adaptation, 71–89, 137–151
   aerobic fitness vs., 76–77, 81
   age and effect of, 140, 149–151
   by Albertson, 71–73, 85, 86
   author's experience of, 117–119, *118*, *119*
   brain as Central Governor for, 76–78, 85, 101
   for cool conditions, 86–87
   endurance physiology understanding and, 75–76
   evolution and, 141–143
   geographical distribution of heat- and cold-related mortality, 143–149, *146*
   heat illness/heatstroke prevention and, 73–75
   heat index and, 138, 141n
   heat-safety guidelines, 128–130
   hot yoga as, 177
   minor changes for, 137–138
   passive heat therapy for, 72, 85, 88
   for performance, 78–81
   for stress resistance, 87–89
   stress resistance and, 52, 87–89
   training for, 81–86
   wet-bulb globe temperature (WBGT) and, 139n, 141, 147
   wet-bulb temperature (WBT) and, 138–141
heat-bathing traditions. *see* Finland, sauna in; hot baths and hot tub use; sauna
Heat Center, Martin Army Community Hospital (Fort Benning), 100
heat dome, 111
heat flux measurement, 82–84
heat.gov, xvi
heat illness and heat stroke, 121–135. *see also* heat adaptation
   Army guidelines and, 122–123
   cold plunging for, 75, 127, 193
   exertional vs. classical (passive) heatstroke, 15n
   football deaths and injuries from, 121–130
   heat exhaustion, 100
   heat-safety guidelines in Georgia, 126–130
   hydration and, 98–100
   involuntary heat exposure and, 131
   minor changes and safety, 137–138
   prevention of, 8–17, 73–75
   types of illnesses, 128
   workers' risk of, 130–135
heat index, 138, 141n
heat-island effect, 64–66
heat shock proteins, 52–54, 80–81
heat tolerance, 3–17
   "Hotter'N Hell Hundred" bicycle marathon and, 3–8, 10, 11, 14
   safety issues of, 6–7, 12, 14–16
   sweating and (*see* sweat)
   testing for, 7–16
   thermal equilibrium (homeostasis) and, 14, 16, 24
   training/rehabilitation programs for, 7, 12–13, 16–17
heat waves (Europe, 2023), 140
The Heat Will Kill You First (book), xvii
Heckel HT3000, 217–219
hemoglobin mass, 86–87
high school football, 121–130, 137–138
Hill, Austin Bradford, 48–50
Hippocrates, 127, 218
Hof, Wim, 191–194, 196
"Hold Your Wee for a Wii" contest, 103n
homeostasis (thermal equilibrium), 14, 16, 24
Homo sapiens, 57

INDEX

Hoover Dam, 131
Hopp, Jeff, 127–128, 129
hot baths and hot tub use
  benefits of, 55–56, 151
  as passive heat therapy, 72, 85, 88
  Roman baths, history of, *163*, 163–167, 191
Hotter'N Hell Hundred bicycle marathon, 111–119
  author's heat adaptation and endurance in, 117–119, *118*, *119*
  event (2023), 4–8, 10, 111–117, 234
  heat tolerance and, 3–8, 10, 11, 14
  training for, 68
HOTWORX, 179–181
hot yoga, 175–181, 211–212, 224n
Hubbard, L. Ron, 181, 183
Huber, Robert, 138
Huggins, Robert ("Huggy Bear")
  field work at road races, 74, 75
  on football safety issues, 124, 132
  heat acclimation training, 80–82, 91–92, 105, 108, 118
  laboratory research of, 7–11, 15–16, 68
humidity
  sweat evaporation and, 74
  wet-bulb globe temperature (WBGT), 139n, 141, 147
  wet-bulb temperature (WBT), 138–141
hydration, 91–110
  CamelBak invention for, 114–115
  controversy of, 92–93
  dehydration symptoms, 100–101, 160
  8-by-8 controversy, 105
  heat illness and, 100
  hyponatremia from, 93–96, 98, 99, 102–104
  marketing by beverage industry and, 104–106
  mild hydration and performance, 100–103
  recommendations for, 96–97, 102n, 106–107
  research by US military on, 98–100, 107
  research funding and, 97–98
  sweat rate calculation and, 108–109
  Sweat Test and, 91–92, 102, 107
  thirst and, 93, 95, 105–107
  types of beverages for, 109–110
hydrofluorocarbons (HFCs), 63
hydrotherapy, 191
hyperthermia for depression, 212–213, 214–219, 222, 223, 229. *see also* CHILL'D (Cold and Heat to Investigate Lowered Levels of Depression) study
hyponatremia, 93–96, 98, 99, 102–104

ice, early inventions, 59–61, *60*
ice bath, for heatstroke, 75, 127, 193
Iden, Gustav, 79
immunotherapy, 218
India
  air-conditioning in, 62
  climate change and, 63
  early ice use in, 59
inflammation, 195, 208, 215–218, 221
infrared sauna, 5, 54, 179, 203, 219. *see also* CHILL'D (Cold and Heat to Investigate Lowered Levels of Depression) study
International Association of Firefighters, 185
interoception, 221
Inuit people, 142–143
Iran, early climate control, 61–62
Ironman, 95, 100
Israel, 11, 79

JAMA Internal Medicine journal, 44
JAMA Psychiatry, 217
James, LeBron, 191, 194
Japan
  "Cool Biz" initiative, 66
  heat adaptation study, 150
  heat study in, 55–56
  totonou (tidied up), 202, 237
Jefferson, Thomas, 191

jewelry (prehistoric) study
    (Netherlands), 230
Johnson, Bryan, 165
Jones, Forrest, 123–124, 127, 129–130
Jones, Glenn, 123
Journal of Physiology, 105
The Joy of Sweat (Everts), 184–185
Julius, David, 29n
Juntunen, Justin, 187–188

Kalahari Desert (Botswana), 36n
Kamberov, Yana, 32–34
Kaplan, Bob, xviii
Kellogg, John Harvey, 191
Kennefick, Robert, 107
kidneys, 133, 183
Kilpi, Eero, 156–157, 202
Kipruto, Benson, 71
Kirkham, Jeff, 235, 236, 237
Kodawari, 200–202
Koren, Leonard, 190
Korey Stringer Institute (KSI, University
    of Connecticut). see also Casa,
    Douglas; Huggins, Robert
  heat training by, 74, 80, 117–119,
    118, *119*
  namesake of, 123–124, 126, 148
  on reporting of heat-related
    illnesses, 132
  research and testing by, 7–16, 68
  Sweat Test by, 91–92, 102, 107
Kressy, Megan, 155
Kuopio Ischemic Heart Disease Risk
    Factor Study (KIHD), 43–51

lämpömassa (heat mass), 40
Lancet journal, 144
Langan, Sean, 8, 91, 118, *118*, *119*
Laukkanen, Jari, 43–51
Laukkanen, Tanjanina, 44
Laurencin, Isaiah, 123
Laursen, Paul, 101, 102, 106
Lee, Earric, 46, 196, 207
Lembke, Alex, xi–xiv, *168*, *169*, 169–
    172, 173

Le Pen, Marine, 64
Levick, James J., 15
Lieberman, Daniel, xix, 31–32, 35, 36
life expectancy (Finland), 43
Lindley, Chris, 206–207, 211, 227
liver, 53–54, 183
London, England, 145–146, *146*
Lowry, Christopher, 215–216, 221–
    222
Löyly sauna/restaurant (Helsinki
    Harbor), 157–160, 167
löyly (steam), xii–xiv, 40–41, 50, 158,
    168, 169–170
Lu, Yi-Chuan, 141n
Lucero, Cynthia, 93–95, 98, 102,
    103, 105

m. Vaccae, 215
Magic and Mystery in Tibet (David-
    Néel), 213
mammals, body cooling by, 25–26, 31.
    see also sweat
Manhattan '45 (Morris), 234–235
Mantz, Conner, 85
Marietta (Georgia) Blue Devils, 126
Maron, Michael, 24–25
Martin, David, 8, 27, 82, 83, 87, 91,
    114, 118, *118*, *119*
Mason, Ashley, 203–205, 216, *216*,
    218–219, 222, 223, 229
Max (Turtle Boy), 116
McBride, Nia, 11
McDougall, Christopher, 36
McGraw, Martha, 157–161, 172,
    187, 190
Melbourne marathon, 96
mental health, 203–225
  antidepressants for, 105, 206n, 210–
    212, 216–217
  cold plunging for, 196, 198
  depression, "CHILL'D" clinical trial
    for, 203–204, 206–207, 211, *216*,
    216–219, 222–225, 227–228
  depression and body temperature,
    204–206, *205*

depression and hot yoga, 211–212, 224n
hyperthermia for depression, Raison on, 212–213, 214–219, 222, 223, 229
  inflammation and fever for, 215–218, 221
  interoception and, 221
  severe depression, defined, 222–223
  social benefits of cold plunging, 199–200
  social benefits of sauna and heat rituals, 156–157, 171–173, 229–237
  Tummo for, 213–214
  warm physical sensation and positivity, 219–221
metabolic heat, monitoring of, 10–12, 16
Miami Dolphins, 148
mice, 33, 88, 195
military research and practices, 11, 79, 98–100, 107, 122–123, 129–130
Miller, Arthur, 58
Mind-Body Institute (Harvard), 213
Minnesota, heat-safety guidelines of, 132
Minnesota Vikings, 147–148
Minson, Christopher, 29, 31, 35, 55, 56, 74, 83n, 86, 88–89, 96, 119
Mirkin, Gabe, 195
mitochondria, 142–143
Mohammed (prophet), 165
Morgan, Rich, 126–127, 128–129
Morris, Jan, 234–235
mortality risk and temperature, 143–149, *146. see also* heat adaptation; heat illness
MountainStrong (Vail Behavioral Health), 209–211
Muscogee Creek Native Americans, 20
Mustonen, Kim, 161–162

Narconon (Church of Scientology), 181
NASA, 40, 64

National Institute of Occupational Safety and Health, 131
NCAA, 124, 128–129
Neanderthals, 34, 141, 230n
nematode worms, 54
nervous system. *see also* mental health
  heart health and, 207
  nociceptors, 189
  parasympathetic, 192, 207
  sympathetic, 29, 189, 192
Netherlands, 197–198, 230
Newton, Isaac, 22, 22n, 28
NFL, 124
Nicaragua, 134
Nix device, 108–109
Noakes, Timothy, 75–76, 94–98, 100
nociceptors, 189
North American Sauna Society, 156

Obama, Barack, 131
Occupational Safety and Health Administration (OSHA), 131–132
Olympic Games
  Team Trials (2020), 71
  Team Trials (2024), 77
  Tokyo (2020/2021), 78–79
Oregon, 132
Othership, 231–234
Outlive (Attia and Gifford), xviii, 109n
Oxford University, 160

Pääkkönen, Jasper, 157
Pannier-Runacher, Agnès, 64
Panorama Landscape Hotel, 161–162
parasympathetic nervous system, 192, 207
passive heat therapy, 72, 85, 88
Patrick, Rhonda, 72
Pedialyte, 109
Pennsylvania State University (Penn State), 9, 126, 137, 140
PepsiCo, 97
Persian Gulf nations, 133–134
persistence hunting hypothesis, 35–36
Phelps, Michael, 191

physiology and heat
  air-conditioning and, 57–68 (*see also* air-conditioning)
  endurance physiology, understanding of, 75–76
  environmental physiology, 45
  health benefits, 39–56 (*see also* health benefits of heat)
  heat tolerance and, 3–17 (*see also* heat tolerance)
  sweating and, 19–37 (*see also* sweat)
Planet Beach, 179
polycystic ovarian syndrome (PCOS), 53
positivity, warm physical sensation and, 219–221
poverty, heat and, 9
Power, Matthew, 144
Powerade, 104
power of heat. *see also* endurance; health benefits of heat; heat tolerance; physiology and heat; sweat
  car analogy and, 10, 24
  climate change and, xvi–xvii
  feelings evoked by, xv–xvi, xvii
  healing vs. danger of, xvii
  sweat and its benefits, xvii–xxi
Proceedings of the National Academy of Sciences (*PNAS*), 138, 139
Prozac, 210–211
Pruitt, Robby, 123
Purification Rundown (Church of Scientology), 181–182
Purple Hurricanes (Fitzgerald, Georgia High School football), 121–124
Püustinen, Anu, 158–159, 167

Qatar, 133

Raimondo, Gina, xvi
Raison, Charles, 212–213, 214–215, 216–219, 222, 223, 227, 229
Rajaportti sauna (Tampere, Finland), xi–xiv, xiii, 157, 162–163, 167–173, *168*, *169*

rats, 215
rectal probe/thermometer, 10–12, 80–81
The Relaxation Response (Benson), 213
Repasky, Elizabeth, 24n
research. *see also* health benefits of heat; *individual names of research sources and studies*
  double-blind studies, 217
  epidemiological research on cause and effect, 47–49
  healthy user bias, 46
resting body temperature, 118
RICE (rest, ice, compression, elevation), 195
Ritossa, Ferruccio, 53n
Rollins, Henry, xv–xvi
Roman baths, history of, *163*, 163–167, 191
Romps, David, 141
Roosevelt, Franklin D., 61
Royal College of Physicians, 19
Royal Society, 20
runners and running, xxi, 24, 71–78, 93–96, 100–102, 116
Russia, *banya* in, 55, 161

safety issues. *see* cold plunging; heat illness and heatstroke
Sahara Desert, 141
Salazar, Alberto, 74
salinity, sweat, 85
sauna. *see also* CHILL'D (Cold and Heat to Investigate Lowered Levels of Depression) study; Finland, sauna in
  after workouts, 85
  Finnish practice of (*see* Finland)
  for heat adaptation, 149
  infrared, 179, 219
  infrared sauna, 203
  lämpömassa (heat mass), 40
  research on benefits of, xvii–xix
  "sauna drunk," 202, 237
  slowing down as goal of, 156, 190
  US popularity of, 155–156

Sauna: The Finnish Bath (Viherjuuri), 167
"Sauna Bro" stereotype, 155
Sauna Days, 187, 188, 198–199
Sauna Society (*Suomalainen Saunaseura*), 39–42
Sauna Times blog, 199
savusauna (smoke sauna), 158
Scott, Annette, 201
Scott, John A., 113n
Searcy, Carlton, 121, 122–123, 125
Searcy, Don'terio Jacquavious ("DJ"), 121–124, 126–130, *130*
Searcy, Jaclyn, 122, 123
secretory coil, 27
self-perceived heat-stress level, 83n
serotonin, 216
Shell, 104
Sherwood, Steven, 138
Shields, Brooke, 93, 103
"shotgun shacks," 61
siesta, 142
Simpson, Tom, 77
sisu (grit), 41, 172, 216
skin blood flow, 28, 50–51, 73–74
skin temperature, 28, 82, 85, 198
Smith, Stephen P., 179
social connection. *see also* mental health
  cold plunging for, 199–200
  sauna and heat rituals for, 156–157, 171–173, 229–237
Sorensen, Chloe, 217, 223
South Africa, 12–13, 95
Spain, 166
Spelman College, 64
sperm, sauna and, 165
sports science, inception of, 13. *see also* endurance; health benefits of heat; physiology and heat
SSRI antidepressants, 105, 210–212, 216–217
states, heat-safety guidelines of, 132
Stornes, Casper, 79
The Story of the Human Body (Lieberman), xix

stress resistance, heat training for, 52, 87–89
Stringer, Kelci, 148n
Stringer, Korey, 123–124, 126, 148
Summers, Ashley Miller, 103
Sunday Times (London), 192
sweat, 19–37. *see also* detoxification; hydration
  antidepressants and, 206
  benefits of, xvii–xxi
  DNA study of prehistoric artifact, 230–231
  evaporation of, 29–32, *30*, 74
  evolution of sweating in humans, 32–37
  heat adaptation and, 150
  indoor lifestyle and lack of, 66–68
  physiology and external heat, early understanding, 19–23
  salinity of, 85
  from sauna, 41
  sweat rate calculation methods, 91–92, 102, 107, 108–109
  thermoregulation and, 23–26, 34, 36–37
  typical rate of, 31
Sweat (Aaland), xviii–xix, 165–166
sweat lodge example, 235–237
Sweat Program, 181
Sweat Test, 91–92, 102, 107
Swoboda, Ryan, xxi, 14–17, 25, 102, 125
sympathetic (fight-or-flight) nervous system, 29
sympathetic nervous system, 29, 189, 192

Tampere, Finland. *see* Rajaportti sauna
Teamsters, 131
temperature receptor pathways (TRPs), 28
temperature-sensing capsule, 10, 16
TempPredict, 204–205, *205*
Texas, 133. *see also* Hotter'N Hell Hundred bicycle marathon
Texas Monthly magazine, 6

INDEX

Texas Turtles, 234
Thailand, 145–146, *146*
Thermae of Caracalla, 164, 235
thermal equilibrium (homeostasis), 14, 16, 24
Therma Bucharest, 235
thermoregulation. *see also* sweat
  behavioral, 25
  blood circulation and skin, 28, 50–51, 73–74
  cold water immersion and, 200
  defined, 24n
  importance of, 23–26
  sweating for, 26, 34, 36–37
  thermoneutral status and, 27
thirst, 93, 95, 105–107. *see also* hydration
"three-a-day" (football), 121
TikTok, 93, 105
The Times (London), 104
Timmerman, Mark, 46
Tipton, Mike, 189, 192–193, 198, 200
totonou (tidied up), 202, 237
Tour de France, 77, 78, 83, 96, 109, 191
training/rehabilitation programs. *see* heat adaptation
transcription factors, 33n
trees, 64–65
Trump, Donald, 132
Tucker, Ross, 76
Tudor, Frederic ("Ice King"), 59
Tummo, 213–214
Turtle Boys (Texas Turtles), 115–117
"two-a-day" (football), 128
Tykkynen, Marjo, 171

UCLA, 220
Ukraine, *banya* in, 161
uncompensable conditions, 9–10, 116
Undesigning the Bath (Koren), 190
The Uninhabitable Earth, 139
United Nations, 160
United States
  American sauna culture, 187–189, 198–202
  heat-safety federal guidelines, 131–132, 133–134
  military research and practices, 79, 98–100, 107, 122–123, 129–130
University of Alabama, 125
University of Central Florida, 17
University of Connecticut. *see* Korey Stringer Institute; Korey Stringer Institute (KSI)
University of Loughborough, 104
University of Virginia, 14–15, 17
UPS, 130–131
USA Track & Field, 102

Vail Behavioral Health, 209–211
Valtin, Heinz, 105–106
vasopressin, 105n
Velotron bike, 9, 10, 80
Viherjuuri, H.J., 167
VO_2 max, 76, 150

Wallace, Doug, 142
Waon therapy, 55
Washington, heat-safety guidelines of, 132
Waterlogged (Noakes), 98
wet-bulb globe temperature (WBGT), 139n, 141, 147
wet-bulb temperature (WBT), 138–141
What Doesn't Kill Us (Carney), 192
Williams, Lawrence, 220
Wim Hof Method, 191–194, 196
Wire, Mike, 182
Wood, Levison, 144
worker safety, heat illness and, 130–135, 149
World Happiness Report, 160
Wright, Frank Lloyd, 62, 63
Wright, Lawrence, 6

Yegül, Fikret, 165
yoga, hot, 175–181, 211–212, 224n
Yoga College of India, 175
Young, Clayton, 85